T0339612

HIGHWAY BRIDGE MAINTENANCE PLANNING AND SCHEDULING

HIGHWAY BRIDGE MAINTENANCE PLANNING AND SCHEDULING

MARK HURT

Bridge Design, Bureau of Structures and Geotechnical Services, Kansas Department of Transportation

STEVEN D. SCHROCK

Department of Civil, Environmental, and Architectural Engineering, University of Kansas

Amsterdam • Boston • Heidelberg • London
New York • Oxford • Paris • San Diego
San Francisco • Singapore • Sydney • Tokyo

Butterworth-Heinemann is an imprint of Elsevier

Butterworth-Heinemann is an imprint of Elsevier
The Boulevard, Langford Lane, Kidlington, Oxford OX5 1GB, UK
50 Hampshire Street, 5th Floor, Cambridge, MA 02139, USA

Notices
Knowledge and best practice in this field are constantly changing. As new research and experience
broaden our understanding, changes in research methods, professional practices, or medical treatment
may become necessary.

Practitioners and researchers must always rely on their own experience and knowledge in evaluating and
using any information, methods, compounds, or experiments described herein. In using such information or methods they should be mindful of their own safety and the safety of others, including parties
for whom they have a professional responsibility.

To the fullest extent of the law, neither the Publisher nor the authors, contributors, or editors, assume
any liability for any injury and/or damage to persons or property as a matter of products liability,
negligence or otherwise, or from any use or operation of any methods, products, instructions, or ideas
contained in the material herein.

Library of Congress Cataloging-in-Publication Data
A catalog record for this book is available from the Library of Congress

British Library Cataloguing-in-Publication Data
A catalogue record for this book is available from the British Library

ISBN: 978-0-12-802069-2

For information on all Butterworth-Heinemann publications
visit our website at http://www.elsevier.com/

Working together
to grow libraries in
developing countries

www.elsevier.com • www.bookaid.org

CONTENTS

Acknowledgment   *ix*

1. Introduction   **1**
    1.1 Bridges in the United States   1
    1.2 Bridge Preservation Process   5
    1.3 Bridge Preservation Practices Before 1970   8
    1.4 Development of the National Bridge Inspection Standards   13
    1.5 Ongoing Changes in Practice   19
        1.5.1 Collapse of the I-35W Bridge   19
        1.5.2 Preventing the Next Failure   23
        1.5.3 Cost of Failure   26
    1.6 Managing the Bridge Preservation Process   27
    1.7 Scope and Purpose of the Text   29
    References   29

2. Bridge Elements and Materials   **31**
    2.1 Classification of Bridge Structures   31
    2.2 Buried Structures   32
    2.3 Elements of Span Bridge Structures   38
        2.3.1 Substructure Elements   39
        2.3.2 Superstructure Types and Elements   47
    2.4 Bridge Mechanics   65
        2.4.1 Axial Forces – Tension   66
        2.4.2 Axial Forces – Compression   68
        2.4.3 Forces From Transverse Loading – Bending   70
        2.4.4 Transverse Loadings – Shear   72
        2.4.5 Fracture and Fatigue   74
    2.5 Bridge Materials   77
        2.5.1 Concrete and Reinforced Concrete   77
        2.5.2 Prestressed Concrete   84
        2.5.3 Steel   86
        2.5.4 Timber   92
        2.5.5 Other Materials   96
    References   98

3. Bridge Inspection and Evaluation   **99**
    3.1 Introduction   99
    3.2 Bridge Inspection in the United States   100

3.2.1 Types of Inspection   100
3.2.2 Component Condition Ratings   105
3.2.3 Appraisal Ratings   110
3.2.4 Deficiency and Sufficiency   112
3.2.5 Critical Findings   113
3.2.6 Element-Level Inspection   114
3.3 Bridge Inspections in Canada, Western Europe and South Africa   121
3.3.1 Canada   121
3.3.2 United Kingdom   123
3.3.3 South Africa   124
3.3.4 France   124
3.3.5 Germany   126
3.3.6 Finland   127
3.3.7 Observations   129
3.4 Reliability-Based Bridge Inspection   129
3.4.1 Risk-Based Assessment   129
3.4.2 Implementation in Indiana   131
3.5 Inspection Techniques and Technologies   134
3.5.1 Visual Inspections and Sounding   134
3.5.2 Nondestructive Testing – Concrete   135
3.5.3 Nondestructive Testing – Steel   136
3.5.4 Sampling   137
3.5.5 Inspecting Fiber-Reinforced Polymer   138
3.5.6 Posttensioning Ducts   139
3.5.7 Structural Health Monitoring   140
3.5.8 Sonar and Underwater Inspection   142
3.6 Load Rating   143
3.6.1 General Approach   144
3.6.2 Analysis Methodologies   144
3.6.3 Truck Loadings   148
3.6.4 Load Rating by Testing   150
3.6.5 Fatigue Evaluation of Steel Bridges   151
3.6.6 Programming Maintenance Actions   152
References   153

4. **Preventative Maintenance**   **155**
4.1 Introduction   155
4.2 Cost Effectiveness   156
4.3 Maintenance Inspections   158
4.4 Bridge Decks and Expansion Joints   160
4.4.1 Deck Drainage   160
4.4.2 Deck Patching   162
4.4.3 Deck and Crack Sealing   165

    4.4.4 Deck and Expansion Joint Washing   167
    4.4.5 Timber Deck Preservation   168
  4.5 Bridge Superstructure and Substructure   168
    4.5.1 Washing Superstructures   168
    4.5.2 Sealing Bearing Seats   169
    4.5.3 Bearing Device Maintenance   170
    4.5.4 Painting   170
  4.6 Bridge Substructure and Waterway   172
  4.7 Approaches and Roadways   173
    4.7.1 Driver Guidance   173
    4.7.2 Approach Settlement   176
    4.7.3 Relief Joints   178
  4.8 Recommendations   178
  References   179

5. **Substantial Maintenance and Rehabilitation**   **181**
  5.1 Introduction   181
  5.2 Assessment and Scoping   182
    5.2.1 Closing or Removing Bridges   182
    5.2.2 Level of Repair   183
    5.2.3 Design Codes and Specifications   184
  5.3 Repair Methods   185
    5.3.1 Concrete   185
    5.3.2 Steel   194
  5.4 Substantial Maintenance Actions   213
    5.4.1 Decks and Railing   214
    5.4.2 Expansion Joints   221
    5.4.3 Bearing Devices   226
    5.4.4 Steel Superstructure   231
    5.4.5 Painting Steel Structures   236
    5.4.6 Reinforced Concrete Superstructure   237
    5.4.7 Prestressed Concrete Superstructure   240
    5.4.8 Posttensioned Concrete Superstructure   242
    5.4.9 Piers and Pier Bents   243
    5.4.10 Abutments   247
    5.4.11 Culverts   249
  5.5 Rehabilitation Actions   251
    5.5.1 Deck Replacement   251
    5.5.2 Superstructure Replacement   253
    5.5.3 Bridge Widening   253
  References   253

6. Bridge Life Cycle Costing   **255**
  6.1 Project Scoping and Selection   255
      6.1.1 Initial Selection   255
      6.1.2 Adjusting Scope   257
  6.2 Bridge Life Cycle Cost Analysis   259
      6.2.1 Inflation   259
      6.2.2 Discount Rate   262
      6.2.3 Cash Flow Diagrams   263
      6.2.4 Residual Value   265
  6.3 Determining Costs   266
      6.3.1 Agency Costs   266
      6.3.2 User Costs   270
      6.3.3 Vulnerability Costs   275
      6.3.4 Economic Costs   275
  6.4 Deterioration Rates   275
  6.5 Applying Bridge Life Cycle Costing   282
      6.5.1 BLCCA Programs   282
      6.5.2 Typical Applications of BLCCA   283
  References   286

7. Bridge Management   **289**
  7.1 Contemporary History of Bridge Management   289
      7.1.1 Before the NBIS   289
      7.1.2 Definition of Bridge Management   291
      7.1.3 Level of Service   293
      7.1.4 BMS Software   294
      7.1.5 Asset Management   296
  7.2 Project and Program Selection   298
      7.2.1 Priority Ranking   298
      7.2.2 Programming Maintenance Work   299
      7.2.3 Priority Indexing   302
  7.3 Experiences from the Kansas Department of Transportation   308
      7.3.1 Commitment to Bridge Preservation   308
      7.3.2 Coordination Between Bridge Inspection and Design   309
  References   309

**Appendix 1: Delay Calculation for Undercapacity Flow at a Typical
               Signalized Work Zone for Bridge Deck Repair Work**   **311**
**Appendix 2: Economic Impact Analysis of Deferring Maintenance
               on K-10 Bridges near Desoto, Kansas**   **321**
Subject Index   **325**

ACKNOWLEDGMENT

The authors would like to gratefully acknowledge that figures used in the text are not attributed to specific sources, and which are not their own, but have been graciously provided for use with permission from the Kansas Department of Transportation.

CHAPTER 1

Introduction

"A bridge is to a road as a diamond is to a ring."

– Anonymous

Overview

The inventory of bridges on public roads in the United States is discussed. The bridge preservation process is introduced. Bridge preservation practices before the establishment of the National Bridge Inspection Standards (NBIS) are examined through the case of such practice in the state of Kansas. The development of the NBIS is presented. The nature of ongoing development in bridge inspection and evaluation practices is illustrated by the case of the I-35W Bridge collapse. The implementation of bridge management systems (BMS) to manage bridge inspection and condition data, and the growth of BMS are discussed. An overview of the layout of the book is provided.

1.1 BRIDGES IN THE UNITED STATES

The previous statement may appeal to the vanity of those who work with bridges, but it also reflects a truth: bridges are critical assets that provide important value, but at a cost. Value, in that each highway bridge is a solution to a problem of how to carry traffic across a river or gorge or other obstacle such as conflicting lanes of traffic. Cost, in that the solution comes at a price, each section of a bridge deck costs several times more than an equivalent area of roadway both to construct and to maintain over the life of the bridge.

According to the Federal Highway Administration (FHWA), in 2010 the road network of the United States included over 4,083,768 miles of public roads and more than 604,493 bridges [1]. A bridge is defined by the FHWA as, "a structure including supports erected over a depression or an obstruction, such as water, highway, or railway, and having a track or passageway for carrying traffic or other moving loads, and having an opening measured along the center of the roadway of more than 20 feet." (Figure 1.1)

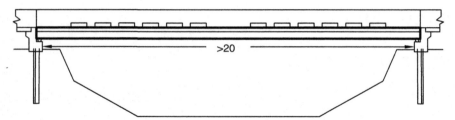

Figure 1.1 *Bridge Opening.*

The total length of those bridges is 16,349.5 miles [2], less than 0.5% of the total miles of public road. The amount expended by all levels of government in the United States in 2010 on public roads and bridges was $205.3 billion. Of this, $60 billion was spent on system rehabilitation, which is defined as, "capital improvements on existing roads and bridges that are intended to preserve the existing pavement and bridge infrastructure." Twenty-eight and half percent, $17.1 billion, of the system rehabilitation expenditures were for bridge-sized structures. This does not include the system rehabilitation funds spent on highway structures with an opening of 20 ft. or less. These small spans and culvert structures are used most often to convey drainage or sometimes to provide a single lane underpass through a roadway berm. These structures are more numerous than bridge-sized structures and are subject to most of the same maintenance issues as the larger structures. The cost of work on structures with an opening of less than 20 ft. conducted under system rehabilitation projects is captured in the $42.9 billion in highway expenditures. Rehabilitating structures for preservation is considerably more expensive than rehabilitating an equal length of roadway.

Part of the cost of bridges is also the acceptance of risk. A study of bridge failures in the United States over the period of 1989–2000, by Wardhana and Hadipriono of Ohio State University, found cases of 503 failures [3]. Failure was defined as the incapacity of a bridge or its components to perform as specified in the design and construction requirements. Conditions of either collapse (total or partial) or distress constitute failure of a bridge and result in its removal from service until either repair or replacement. A distressed bridge is one with one or more components in such condition that the facility is rendered unserviceable. An example would be excessive deflections in the superstructure resulting in a dip in the bridge deck that would render the bridge unusable for traffic. Almost all of the failures with identified conditions were either partial or total collapse. The consequences of the collapse of a bridge can be quite severe and, in the worst case, result in fatalities. In the cases studied, there were 76 fatalities and 161 people injured.

To characterize a bridge as having failed in the study, it was not only implied that it became unserviceable, but that it became unserviceable suddenly and unexpectedly. Of the 503 bridge failures studied by Wardhana and Hadipriono, 266 failed due to high-water events, 103 failed due to either overloading or vehicular impacts, and 45 failed due to other events such as fire or earthquakes. The failures of only 48 bridges were attributed to either deterioration or fatigue. The most common way for a bridge to fail was to be subjected to an extreme event.

A far more common end to the life of a bridge is deterioration that accumulates and results in a progressively less serviceable structure. Under the wear of traffic loads and exposure to the weather and to agents such as salts used to melt snow and ice on roadways, steel corrodes, and concrete cracks and spalls. The wearing surface of the bridge deck may become rough enough to require slowing traffic. The supporting members of the structure may lose enough material that their ability to bear load is reduced, requiring the restriction of heavy trucks from the bridge. Thankfully, slow deterioration rarely results in a sudden failure with the attending risk of injury to bridge users; however, it may still result in significant economic impact by disrupting traffic. This is particularly true for the movement of commercial freight by heavy trucks.

The cost of bridges makes them a significant investment for owners and operators of highways. The risks and consequences of bridge failure require owners and operators to act to maintain their bridges in good repair. In the United States, these actions have come to be classified as *bridge preservation*. The FHWA defines bridge preservation "as actions or strategies that prevent, delay or reduce deterioration of bridges or bridge elements, restore the function of existing bridges, keep bridges in good condition and extend their life" [4].

Bridge preservation has become increasingly important to the owners and operators of highway bridges in the United States due the age and numbers of bridges in their inventories. According to data from the National Bridge Inventory (NBI) maintained by the FHWA, as of 2013 the average age of bridges carrying traffic on public roads in the United States was 43 years [5]. This is due to the rapid expansion of the highway system and public roads in general after World War II. Figure 1.2 shows the decade of construction for bridges on public roads in the United States constructed between 1910 and 2010. For bridges constructed in the post-World War II period and prior to adaption of the American Association of State Highway and Transportation Officials (AASHTO) Load and Resistance Factor Design (LRFD) Bridge Design Specification, the anticipated service life was 50 years.

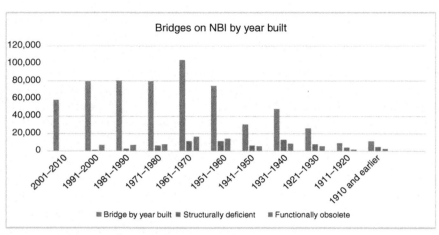

Figure 1.2 *Bridges by Decade of Construction.*

Over 11% of the bridges on the NBI in 2013 are categorized as *structurally deficient* [6]. A bridge is categorized as structurally deficient when one of its major components – the deck, superstructure, or substructure – is rated as poor during a bridge inspection, or when it is evaluated to be inadequate either for load-carrying capacity or for its waterway opening. Structural deficiency does not automatically imply an imminent danger to the traveling public using the bridge. It does imply impairment to the operation of the bridge in that some heavy truck traffic will not be allowed to use the bridge. And it also implies that work is required to restore the condition of the bridge. The poor rating of one or more bridge components is almost always due to deterioration. Deterioration comes about as a function of environmental exposure and use over time. The average age of a bridge rated structurally deficient was 65 years old. By 2023, one in four of the existing bridges in the inventory will be 65 years or older if left in service [7]. For these bridges, preservation actions will be required to maintain them in full service.

A bridge on the NBI may be also considered deficient if it is *functionally obsolete* [6]. A bridge is categorized as functionally obsolete if either the geometry of its deck, the clearance for roadways under the bridge, or the width of the roadway at the approaches to the bridge deck are inadequate. It may also be considered functionally obsolete if either its load-carrying capacity or waterway opening is inadequate, but not to the degree to be considered structurally deficient. If a bridge qualifies as structurally deficient, it is not also considered functionally obsolete. Almost 13% of bridges on the

2010 NBI were functionally obsolete. The total number of deficient bridges in the 2010 NBI was 146,636, over 24% of the total inventory.

1.2 BRIDGE PRESERVATION PROCESS

The term bridge preservation should not be taken to focus solely on the particular maintenance actions to keep a bridge in good condition. These are actions such as sealing open cracks on a bridge deck. Implementing these actions and developing effective strategies for their deployment requires owners and operators to assess the condition of the components of the bridge and to know the relevance of any defects found. Bridge preservation may be defined as a process consisting of three general activities: inspection, evaluation, and maintenance (Figure 1.3).

The cornerstone of the bridge preservation process is inspection. Inspection provides information as to the physical condition of bridge components. An initial inspection provides a baseline for review throughout the life of the structure. Subsequent inspections alert the owner to changes in condition and to any current needs. Maintaining records of bridge inspections allows an owner to track deterioration. Combining the information available from the records of an inventory of bridges over time allows the owner to intelligently predict rates of deterioration and anticipate future needs. Inspection procedures and the intervals at which inspections are conducted are determined by policies adopted by the bridge owner. For bridges on the NBI, those policies are set forth in the NBIS.

Evaluation is an assessment of a bridge's ability to safely carry traffic. Bridge inspectors evaluate the condition of bridge elements during the inspection process. The evaluation step in the bridge preservation process

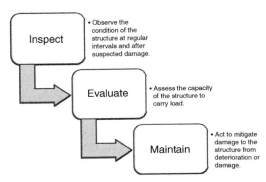

Figure 1.3 *Bridge Preservation Process.*

is an assessment of the bridge as a whole. Bridge owners must determine whether a bridge is safe to remain open after experiencing an extreme event, such as a large flood or a fire. Although if a significant amount of damage is apparent it may be obvious that a bridge needs to be removed from service, often an engineering analysis is required to determine the degree of impairment suffered by the structure. An engineering analysis may also be required to assess the effect of a change in site conditions, such as experienced from stream degradation.

A structural analysis conducted to determine the load–carrying capacity of the existing bridge components in their current condition, noting any loss in capacity due to deterioration or damage, is a load rating. Older bridges may have lower load–carrying capacity than desired for the highway route they service not only due to the effects of deterioration, but the loading used for their initial design may have been significantly less than current standards require.

The current AASHTO LRFD design truck is the HL-93, a 72,000 pound truck with a maximum axle load of 32,000 pounds. Its load effects are combined concurrently with those of a uniform load of 640 pounds per ft. per lane. It was not until 1944 that the design specifications of the predecessor to AASHTO, the American Association of State Highway Officials (AASHO), recommended a minimum design truck load for highways with heavy truck traffic, the H15-S12. The H15-S12 loading consisted of checking for the effects of either a 54,000 pound truck or a 480 pound per ft. lane load with a 13,500 pound concentrated load. This was still considerably heavier than the first weight limits for trucks on public roads in the United States. These were enacted by four states in 1913: 18,000 pounds gross vehicular weight (GVW) in Maine; 24,000 pounds GVW in Pennsylvania and Washington; and 28,000 pounds GVW in Massachusetts [8] (Figures 1.4 and 1.5).

All bridges on the NBI are required by the NBIS to be load rated for the HL-93 truck configuration, note that the concurrent lane loading is not used [9]. The HL-93 truck configuration is known as HS20-44 truck configuration in previous design specifications. Two load ratings are reported to the FHWA: operating and inventory ratings. The operating rating is the maximum permissible weight of truck in the chosen load configuration to which the bridge may be subjected. The inventory rating is the maximum permissible weight of truck in the chosen load configuration, which may safely utilize the bridge for an indefinite period of time. For example, an inventory rating of 39 tons for the HL-93 truck configuration would imply that a truck weighing 39 tons with axles spaced and apportioned similar to the HL-93 should be able to use the bridge indefinitely without

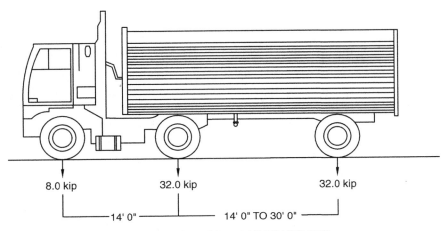

Figure 1.4 *HL-93 Design Truck. (Adapted from AASHTO LRFD [10]).*

causing undue distress on the structure. A HL–93 design truck has a front axle weight of 8,000 pounds and two rear axles each with 32,000 pounds. A 39 ton (78,000 pound) truck in the same configuration would have a front axle of 8,666 pounds (78/72 × 8) and two rear axles of 34,667 pounds.

Maintenance consists of those actions to sustain a bridge in operation despite onslaughts by both deterioration and damage. A broad spectrum of actions will fall into this activity, from actions as simple as cleaning the bridge wearing surface, to as involved as the removal and reconstruction of bridge decks. Bridge maintenance actions can be generally categorized as preventative or substantial.

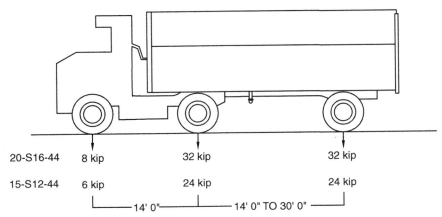

Figure 1.5 *H15-S12 Design Truck. (Adapted from 1941 AASHO Design Manual [11]).*

- Preventative maintenance – actions undertaken to prevent or mitigate deterioration from environmental conditions and/or wear from users.
- Substantial maintenance – actions undertaken to repair damage from deterioration, wear, or traffic crashes.

Generally, preventative maintenance actions require less effort in labor and equipment, while substantial maintenance may require the involvement of engineers to prepare plans and construction contractors to conduct the work. There are also maintenance actions, which overlap the categories, such as sealing a bridge deck wearing surface with a polymer membrane to seal out water and road salts.

The goal of all maintenance work is to keep an existing facility in service. Such work should bring the bridge component addressed to a condition at least equal to the original, as-built condition. Sometimes, lesser criteria may be acceptable as long as the safety of the bridge user is maintained. Work which is done that brings the component addressed to current, or comparable, standards for function may be termed rehabilitation. For example, replacing the deteriorated concrete and reinforcing steel in a bridge deck is a substantial maintenance action. Replacing the deck with a new one wide enough to provide a roadway with shoulders meeting current standards is a rehabilitation of the deck. Note that there are a number of scopes of work in between these two scopes that would address a bridge with a deteriorated deck. Selection of the proper scope is a function of a number of considerations, including available budget and other needs. For this text, scopes of work from repair to rehabilitation will all be considered as substantial maintenance work.

1.3 BRIDGE PRESERVATION PRACTICES BEFORE 1970

An understanding of how practices for bridge preservation have developed from past work helps one appreciate current practice. Although current practices in the United States are dictated by federal policy and regulation, previously bridge preservation was a more local concern. To illustrate practice prior to implementation of the NBIS, the history of bridge preservation in the State of Kansas is examined. To facilitate the agricultural economy of the state, it has maintained one of the largest networks of public roads and bridges in the nation for most of its history.

When the State of Kansas entered the Union on January 29, 1861, it was prohibited from constructing a state owned and maintained system of roads. Its constitution contained a provision that "the state shall never be a party of carrying on any works of internal improvements." Had that section of the state constitution not been amended in 1928 with the proviso "except that"... "It may adopt, construct, reconstruct, and maintain a state system of highways, but no general property tax shall ever be laid nor general obligation bonds issued by the state for such highways" there would have been no state highway system. The reason for the original prohibition was that after the success of the Erie Canal in spurring the economy of New York in the early 1800s, many states financed a number of infrastructure projects (roads, canals, railroads, etc.) with land grants and cash. Many of these projects went bankrupt or were never constructed, with profound negative economic effects on the states in some cases [12].

It was believed by officials of the time that roads were a purely local matter. Roads were constructed only after 12 households within a given vicinity petitioned the county commissioners to have one. The primary mode of travel between cities for people and freight was by rail.

Prior to the advent of automotive traffic, load-capacity requirements for bridges on public roads were relatively light; however, the general condition of the road infrastructure was wanting even for the demands of the time. The first speed limit passed by the legislature in 1869 posed a $5 fine for crossing a bridge at a speed above a walk. As written by a historian, "The state's bridges simply could not take the stress of trot or canter" [12].

At this time, maintenance of the bridges carrying public roads was done by the same local landowners and farmers who were maintaining the public roads. The first comprehensive road law in Kansas was passed by the territorial legislature in January 1860 – a year prior to admission to the Union. It contained provisions requiring the counties to levy a poll tax to fund the maintenance of highways. This was often paid in labor by farmers working out their obligation at the rate of $1.50 per day [12].

By 1900, automobiles began to appear on Kansas roads. That year Kansas was 10th in the nation for automobile ownership with 220 cars. By 1910, there were 10,490 cars and by 1912 there were 30,000 cars. But, even in 1908, there was no highway, which crossed the state from border to border and very few roads extended farther than 20 miles. However, the size of the Kansas road system was one of the largest in the country; its excess of 111,000 miles in 1917 it ranked second only to Texas [13]. At the turn of the twentieth century, there was impetus from several directions to improve the system. Within 3 years after the advent of free rural postal delivery in 1896, the US Post Office determined that it would not deliver on unserviceable roads. In 1890, a Kansas division of the League of American Wheelman (cyclists) formed to press for improvement in roads [12]. A new road act passed in 1901 made road and bridge maintenance the responsibility of the counties rather than the townships and the adjacent landowners; however, it did not forbid the payment of tax by labor. Most bridge maintenance on public roads was still performed by the public, itself.

Involvement of the state with the public road and bridge system began with a road law passed by the 1911 legislature authorizing a state engineer connected with the Extension Department of the State Agricultural College (now Kansas State University) to advise and assist counties with road and bridge work at no expense to them. Later, the need to have a state level organization for roads and bridges to accept the offer of federal aid for highways extended by Congress in 1916 resulted in the general highway law of 1917, establishing the Kansas State Highway Commission (KHC). Its power included: apportioning federal aid to the counties; approving the appointment of county engineers (with the power to remove for incompetency); approving plans for bridges with construction costs exceeding $2000; devising, adopting, and furnishing standard plans and specifications for road and bridge construction; and approving private bridge contracts [13].

The KHC also gathered information concerning public roads and bridges throughout the state. This was the beginning of establishing files for the inventory of state bridges. The KHC made a biennial report of its operations to the legislature. The first one contained documentation of bridge failures in the state for the years 1917 and 1918. Bridge failures were not uncommon; however, they resulted in few injuries to people. There were 42 reported failures in 1917 and 81 in 1918. Several of these were due to deteriorated wooden decks, an issue that could have been addressed with regular bridge maintenance (Figure 1.6).

At its inception, the KHC operated by supporting the work of the counties. Under the 1917 law, bridge inspections and responsibility for maintenance still lay with the county engineer; bridge work under $1000 could still be performed with what was termed lay labor (i.e., farmers working off tax obligations) [13]. Though a state system of routes was established in 1918, this was only the designation of routes, which traversed the state and would be eligible for federal aid. Construction projects were let and maintenance was performed by the counties.

By 1925, the Agriculture Department (the department containing the US Bureau of Public Roads) had stopped approving all new federal funds for work in Kansas due to their dissatisfaction with the output of projects. The state matching appropriations fell far short of what was needed and the situation was compounded by the inadequate staffing and inefficient organization of the commission. The loss of federal funding forced a reorganization and

TABLE No. 15.—BRIDGE FAILURES DURING 1917.

Counties.	Location.	Date.	Type of structures.	Detailed statement of failures of structures.	Amount of damage paid.
Anderson	Ozark Township		Culvert	Horse went through culvert, breaking leg.	$100.00
Anderson	Reeder Township		Culvert	Horse stepped through broken plank in wooden culvert and injured leg.	
Brown	Sec. 26, T. 1, R. 16		Pile bridge	Engine went through 16-ft. pile bridge on south line of section; no one hurt.	51.00
Brown	Mission Township		Bridge	Horse stepped through hole in bridge.	
Brown	Sec. 3, T. 3, R. 17	June, 1917	Concrete bridge	County bridge on township road; 8-ft. concrete arch; 14-ft. roadway; water cut under foundation during flood; county moved 30-ft steel bridge to this location	
Brown	Sec. 7, T. 3, R. 17		Corrugated iron pipe culvert	Corrugated iron pipe culvert, 4 in. in diameter, rebuilt with 6x7 ft. concrete box culvert.	
Butler	Plum Grove Township		Culvert	One culvert washed out.	
Butler	Plum Grove Township		Culvert	Culvert washed in by a truck.	
Butler	Sec. 10-15, T. 25, R. 5		Stone arch bridge	34-ft. span bridge; keystones dropped out; caused by becoming uncovered under heavy traffic; improperly cut stones flattened at haunches; condemned and blown down with dynamite by county commissioners and county engineer; replaced by 20-ft. T girder concrete bridge.	
Cherokee	Salamanca Township		Culvert	Road overseer's horse fell through or off plank culvert that had been condemned, while on inspection of culvert, he thought, hurriedly or carelessly drove on culvert.	
Cherokee	Ryan Bridge	1916	150-ft steel bridge	Ryan Bridge over Spring River taken out by tornado in 1916; rebuilt in 1917.	
Coffey	Pottawatomie		Culvert	Horse valued at $150.00 got leg broken by stepping through hole in culvert.	
Cowley			Concrete slab bridge	Bridge on sand foundation washed out by 10-inch rain or waterspout; the approach was cut out and one abutment undermined causing total collapse.	
Decatur	Sec. 28-29, T. 1, R. 26		Wall	Wall washed out before the fill was put in.	
Decatur	Sappa Township		Walls	Two cement walls washed out.	
Dickinson	Culvert No. 4		Concrete slab culvert	4-ft. span culvert built by a township trustee with only a 6-inch slab and no reinforcement or crushed rock was crushed through by a 10-ton tractor in two places about 18 inches in diameter. The holes were covered with brush by the driver; county engineer hit one of the holes; damaged car to the amount of $60.00.	
Dickinson	Bridge No. .013		Culvert	Stone abutment 10 feet high fell and was replaced by concrete.	
Dickinson	Bridge No. 39			Steel bent set on old boiler tubes; was incased in concrete.	
Dickinson	Bridge No. 47			Condemned for some time and completely overhauled; wing-walls had to be tied together.	
Finney	Bridge at Holcomb		Pile bridge	96 feet washed out replaced with 3 concrete piles to each bent. Two complete bents built with concrete cap and 32-ft. span, using 6 12-inch T beams and 2 13-inch channel for each span.	
Geary	South Milford Road		Culvert	$25.00 paid for skinning up mule which fell through culvert.	25.00
Harper	Chicaskia Township		Bridges	Three small bridges in north part of township which are badly decayed need new floors.	
Harper	Chicaskia Township			New floor needed on county bridge between Sec. 31-32; several holes rotted through.	

Figure 1.6 Partial List of Bridge Failures From the First Report of the KHC. (From first report of Kansas Highway Commission, 1919 [13]).

enlarging of the commission. Also, in 1925, the state passed a tax on gasoline for the purposes of funding roads. Staffing was increased and the commission reorganized into four departments: design, construction, maintenance, and equipment. This was the start in Kansas of the state assuming responsibility for maintenance of public roads and bridges (at least on the State Highway System).

Further reorganization in 1929 resulted in the addition of a maintenance engineer to each of the six field divisions to oversee maintenance activities. For the first time, there was an engineer assigned specifically to be responsible for maintenance work, rather than leaving the activity to the responsibility of a field superintendent with some guidance by the county engineer. From the 1930s onward there was the position of Bridge Maintenance Engineer within the Maintenance Department at headquarters. This engineer was tasked with inspecting the bridge inventory on the State Highway System and assisting maintenance forces in repair and, if necessary, replacement by KHC forces.

The inspections by the KHC Bridge Maintenance Engineer preceded the current paradigm of federally mandated inspections. These inspections, however, were similar in important regards to current practice: the inspection and recording of the condition of bridge elements by experienced engineers with recommendations made regarding needed maintenance work (Figure 1.7).

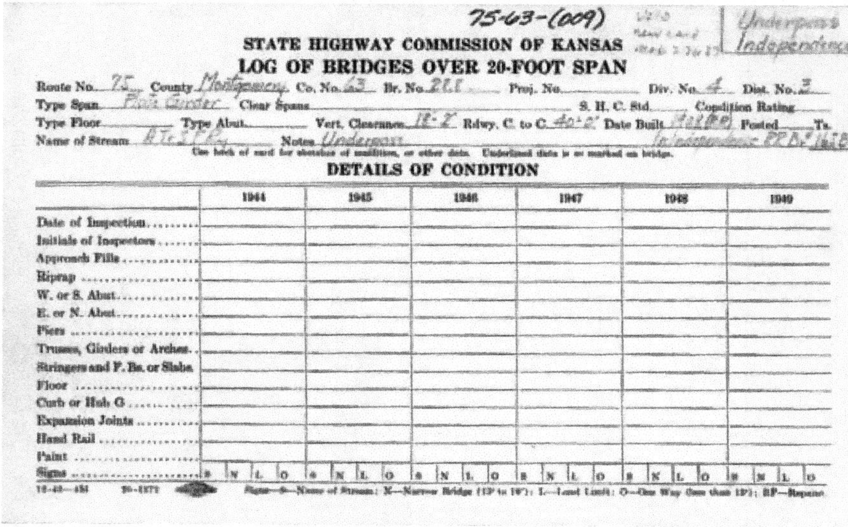

75-63-(009)

Underpass
Independence

STATE HIGHWAY COMMISSION OF KANSAS
LOG OF BRIDGES OVER 20-FOOT SPAN

Route No. 75 County Montgomery Co. No. 63 Br. No. 288 Proj. No. ____ Div. No. 4 Dist. No. 3
Type Span ____ Clear Spans ____ S. H. C. Std. ____ Condition Rating ____
Type Floor ____ Type Abut. ____ Vert. Clearance 18'-2" Rdwy. C. to C. 40'-0" Date Built ____ Posted ____ Ts.
Name of Stream ____ Notes Underpass ____ Independence

DETAILS OF CONDITION

	1944	1945	1946	1947	1948	1949
Date of Inspection						
Initials of Inspectors						
Approach Fills						
Riprap						
W. or S. Abut.						
E. or N. Abut.						
Piers						
Trusses, Girders or Arches						
Stringers and F. Bs. or Slabs						
Floor						
Curb or Hub G.						
Expansion Joints						
Hand Rail						
Paint						
Signs	S N L O	S N L O	S N L O	S N L O	S N L O	S N L O

Figure 1.7 ***Pre-NBIS Kansas Bridge Inspection Form.*** *(From Kansas Department of Transportation (KDOT) files).*

1.4 DEVELOPMENT OF THE NATIONAL BRIDGE INSPECTION STANDARDS

At core of the current bridge preservation practices in the United States is the NBIS. While the driver for the development of road and bridge preservation practices in Kansas after the development of the automobile was the desire to secure federal highway funding, the NBIS standards were created in response to the collapse of the Silver Point Bridge in 1967. Immediately after the collapse, the US Congress, through The Federal Highway Act of 1968, directed the US Secretary of Transportation to establish a NBIS. The NBIS came into effect in 1971 (Figure 1.8).

The Silver Bridge was constructed in 1928 over the Ohio River between Point Pleasant, West Virginia and Gallipolis, Ohio. The main superstructure was a 1760-ft.-long eyebar chain suspension bridge, with a center span of 700 ft. and anchor spans of 380 ft. on either side. At about 5:00 in the evening of December 15, 1967, an eyebar just west of the tower on the west side of the center span suddenly fractured resulting in the immediate collapse of the structure and the loss of 46 people and serious injury to 9 others. The fracture of the eyebar was the result of the growth of a minute imperfection, present from the initial casting of the eyebar, due to stress corrosion cracking [14] (Figure 1.9).

The Silver Bridge had been regularly inspected and maintained prior to the collapse. It had originally been constructed as a private toll facility,

Figure 1.8 *The Silver Bridge (in Foreground). (From 1968 NSTB Report [14]).*

Figure 1.9 *Wreckage of the Silver Bridge on the Ohio Side. (From 1968 NSTB Report [14]).*

was bought by the State of West Virginia in 1941 and finally converted to a nontoll facility in 1951. Prior to this conversion, the bridge was inspected and reports made to the West Virginia Bridge Maintenance Engineer, with subsequent repairs made. According to the West Virginia Department of Transportation the bridge was inspected again in 1959, 1963, 1964, and 1965. However, the defect which brought the bridge down was inside the eyebar and would not have been detectible with technology available at the time (Figure 1.10).

In response to the failure, President Johnson created the President's Task Force on Bridge Safety which was charged, in part, with developing standards to prevent the collapse of other bridges in the United States. Prior to this, bridge inspection and maintenance was seen by the federal government as a purely state and local issue. Indeed, significant federal involvement with public roads outside of those on federal lands had not occurred until the Federal Aid Highway Act of 1916. This disaster was the catalyst to develop national standards for bridges.

The original NBIS in 1971 established the national policies for frequency of bridge inspections, the procedures to be followed, the qualification of the inspection personnel, the format of reporting, and the information required to be maintained for the inventory of each jurisdiction's bridges. States were the jurisdictions responsible for all bridges within their borders, except for bridges owned and operated directly by Federal agencies, such as the Park Service, and bridges on tribal lands. Bridge inspections were

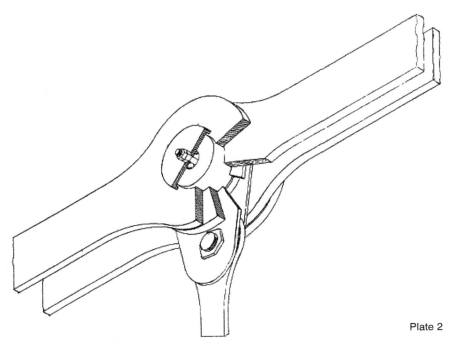

Plate 2

Figure 1.10 *Detail of Eyebar From the National Transportation Safety Board Report.* *(From 1968 NSTB Report [14]).*

to be conducted at least every 2 years, following procedure developed and published by AASHTO. Structure Inventory and Appraisal data were to be collected and retained by the bridge owner and transmitted to the FHWA. As part of these data, each bridge was to be rated for its safe-load-carrying capacity.

The NBIS in 1971 applied only to bridges on Federal Aid highways. This was extended to all public bridges with the Surface Transportation Assistance Act of 1978. The Surface Transportation and Uniform Relocation Assistance Act of 1987 expanded the scope of bridge inspection programs to include two new types of special inspection procedures. Bridges with fracture critical members would receive closer inspection of those member and bridges with foundation members under water would have special inspection of those foundations (Figure 1.11).

The impetus for fracture critical inspections came from the collapse of the Mianus River Bridge in Greenwich, Connecticut in 1983. The Mianus River Bridge was a 2656 ft. long welded steel plate girder bridge built in 1958, which carried Interstate 95. At about 1:30 in the morning on June 28,

Figure 1.11 *Collapsed Section of Mianus River Bridge. (From 1983 NSTB Report [16]).*

1983 a 100 ft. section of the bridge superstructure collapsed into the river below, killing three people and severely injuring three others. As originally constructed, this section of the span was suspended by pin and hanger assemblies. The National Transportation Safety Board (NTSB) determined that an inside steel plate hanger had slipped off its lower pin on the southeast assemblies. Later, the outside hanger of that assembly moved outward from the girder. The increased loading and eccentricity on the upper pin cause a fatigue crack to form which eventually fractured. The failure of the pin led to collapse of the suspended section [15] (Figure 1.12).

In its report, the NTSB cited deficiencies in the bridge maintenance and inspection policies of the Connecticut Department of Transportation as the probable cause of the collapse. Deck drains provided in the original construction had been paved over during maintenance work on the bridge. Runoff from the deck, contaminated in the winter with salts used to melt snow and ice, ran onto the pin and hanger assemblies causing corrosion, which induced distortions in the assembly. Subsequent bridge inspections failed to discover the condition of the pin and hanger assemblies. Current standards for the inspection of fracture critical structures require an inspection of the fracture critical members at no more than arm's length away by

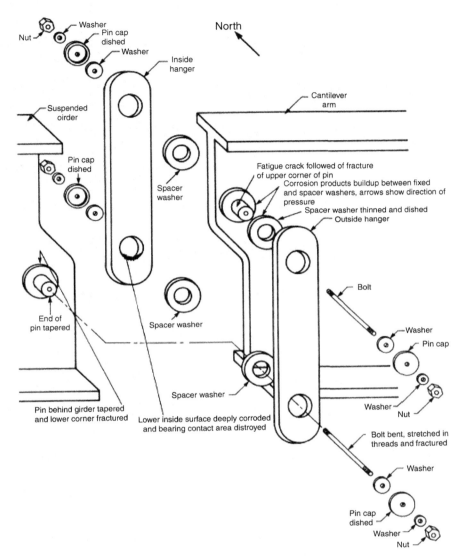

Figure 1.12 *Sequence of Failure in Pin and Hangar Assembly per NTSB. (From 1983 NSTB Report [15]).*

inspectors certified for such work at intervals of no more than 24 months (Figure 1.13).

Requirements for the inspection of the underwater foundation elements of bridges were developed as a result of the failure of the bridge carrying the New York State Thruway over Schoharie Creek in 1987 [16]. After a

Figure 1.13 *Collapse of NY State Thruway Over Schoharie. (From FHWA Underwater Inspection Manual [16]).*

50-year flood event on the creek, the spread footing under one of the river piers was completely undermined by scour action. The footing was founded on dense glacial till, prone to being scoured away and was inadequately protected by rip rap stone. The changes to the NBIS in 1988 required all bridge owners to assess the vulnerability of bridges over waterways to scour damage. The underwater components of bridges are to be inspected at intervals not to exceed 60 months. For bridges deemed scour critical, owners must develop plans of action to be implemented in the case of dangerous flood events. The plans are required to be submitted and filed with FHWA as part of the documentation for the state's bridge inventory.

The last update to the NBIS was published in the Federal Register in December 2004. In addition to clarifying some of the language of the regulation, this update required bridge inspection programs to address quality control and assurance issues to insure the quality of inspection data and load ratings. Programs were also to establish procedures to ensure that critical inspection findings were addressed. The text of the NBIS is listed in Appendix 1.

Current bridge inspection procedures in the United States will be examined in a later chapter to show the information available to engineers planning bridge maintenance activities. This information includes the traditional

NBI reports and the newer bridge element level reports. US bridge inspection requirements will be compared to bridge inspection requirements in selected other countries. The work to develop a risk-based bridge inspection paradigm in the United States will also be discussed.

1.5 ONGOING CHANGES IN PRACTICE

Although the federal regulations comprising the NBIS may only be updated intermittently, the practices involved in inspection and evaluation are adapted as needed to include the latest lessons learned. Any time there is a major bridge failure, the bridge engineering community examines the incident to gain the knowledge to prevent its reoccurrence. As an example of a contemporary collapse that resulted in changes in practice, the collapse of the I-35W Bridge in Minneapolis is reviewed.

1.5.1 Collapse of the I-35W Bridge

On August 1, 2007, approximately 1000 ft. of the deck truss section of the I-35W Bridge in Minneapolis, Minnesota collapsed falling into the Mississippi River and adjacent land below. Thirteen people died and 145 were injured as 111 vehicles went down with collapsed portion of the bridge. In the 10 years prior to this incident, there had been other collapses of Interstate bridges in the United States resulting in fatalities. On October 17, 1989, the Loma Prieta earthquake in Oakland, California resulted in the collapse of the Cypress Street Viaduct carrying I-880 (Nimitz Freeway) and the Oakland Bay Bridge carrying I-80, killing a total of 43 people. On May 26, 2002, a barge struck a pier of the bridge carrying I-40 across the Arkansas River near Webbers Falls, Oklahoma. A 503-ft.-long section of the superstructure fell into the river below resulting in the deaths of 14 people. The collapse in Minnesota, however, was particularly shocking in that unlike the other incidents, there was no obvious external cause. There was no earthquake, no vehicular impact; instead, the I-35W Bridge was in service and carrying commuter traffic at 6:05 pm when the collapse occurred. The only thing out of the ordinary was that a repair project was occurring on the structure at that time (Figure 1.14).

The I-35W Bridge was first opened to traffic in 1967. It carried eight lanes of traffic, four northbound and four southbound, two approximately 52-ft.-wide roadways carried on twin adjacent slabs separated by a longitudinal 6-in. gap running the length of the bridge. The 1907-ft.-long bridge had 14 spans. The majority (1064 ft.) of the superstructure type was a steel

Figure 1.14 *Collapsed I-35W Bridge. (From 2008 NTSB Report [17]).*

deck truss. The adjacent superstructure type was welded steel girder. The average daily traffic, as of traffic information dated 2004, for this section of Interstate 35W was 141,000 automobiles, 5,640 of which were heavy commercial vehicles (Figure 1.15).

The deck of the bridge sat atop 27-in.-deep wide flange rolled steel stringers, running longitudinally at 8 ft. 1 in. on center, attached to transverse

Figure 1.15 *Center Span of I-35W Before Collapse. (From 2008 NSTB Report [17]).*

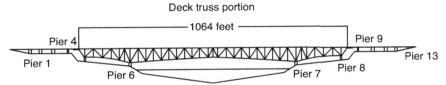

Figure 1.16 *East Elevation of I-35W Bridge. (From 2008 NSTB Report [17]).*

welded floor trusses spaced at 38 ft. on center, which framed into the main deck trusses. The deck trusses were 72 ft. 4 in. apart and spanned pier to pier. Framing the superstructure in this manner resulted in a bridge that is categorized as "fracture critical." A fracture critical structure is one with elements such that the loss of one can result in the collapse of the entire structure. In the case of the I-35W Bridge, the failure of one of the components of a deck truss in the center span led immediately to the collapse (Figure 1.16).

Immediately after the I-35W collapse, the NTSB began an investigation of the incident. The FHWA assisted in that investigation, but it also immediately acted to direct the States and other owners of bridges carrying public roads to take measures to prevent a similar collapse from occurring elsewhere. The day after the collapse, August 2, the FHWA issued Technical Advisory 5140.27, advising the immediate reinspection of all fracture critical deck truss bridges.

The one aspect out of the ordinary immediately prior to the collapse was an ongoing maintenance project on the bridge. The scope of work included removal and replacement of the top 2 in. of the concrete wearing surface. According to estimates made for the NTSB's report on the collapse, the weight of the combination of construction materials, equipment, and personnel staged on the bridge above the site of the initial failure was just under 580,000 pounds. A week after the collapse, on August 8, Technical Advisory 5140.28 was issued to review loadings on existing bridges resulting from construction activities or stocking material for such activities (Figure 1.17).

By the beginning of 2008, the NTSB had already identified the primary cause of the collapse of the I-35W Bridge as an uncaught design error on the part of the designers of the deck trusses. Trusses are assemblies of various members. The primary members are assumed to function in primary tension or compression. The top and bottom members running (primarily) horizontally are chords. The chords are joined by vertical post and diagonal members. The diagonal posts and chords are typically connected via gusset

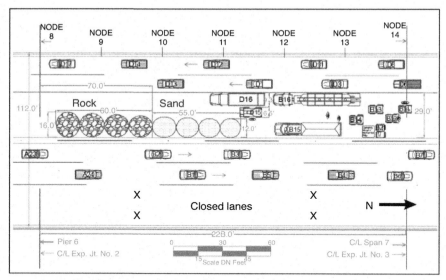

Figure 1.17 *Location of Aggregate and Equipment on I-35W per NTSB. (From 2008 NSTB Report [17]).*

plates. On the I–35W Bridge, each of the main deck trusses was found to have undersized gusset plates at 8 of their 112 nodes (Figure 1.18).

The NTSB reported in its final report that, according to the design methodology and procedures current at the time of the design of the I–35W Bridge, 24 gusset plates should have been 1 in. thick, as opposed to the 0.5-in. thickness specified by the designer. The collapse was initiated by the fracture of the gusset plates at the U10 nodes, a location which according

Figure 1.18 *Connections at Node U10 West. (From 2008 NSTB Report [17]).*

Figure 1.19 *Location of Node U10. (From 2008 NSTB Report [17]).*

to postincident finite element analysis by the FHWA, was beyond yield stress in the originally constructed dead load only condition (Figure 1.19).

1.5.2 Preventing the Next Failure

Although the failure was attributed to design error, the fact remains that this bridge was in continuous service since 1967 and carried traffic for 40 years without incident. This means that bridges currently in service could have similar deficiencies without any obvious symptoms. This failure directly resulted in four actions being mandated by the FHWA for bridges carrying traffic on public roads in the United States:

- Immediate reinspection of all bridges with a superstructure configuration similar to the deck truss on the I–35W Bridge;
- An amendment to bridge construction engineering practices by requiring the review of loadings on in-service bridges from the stockpiling of construction materials;
- Directing the attention of bridge inspectors to possible signs of buckling in gusset plates; and
- Adding a gusset plate review to the load-rating process of inventoried structures.

Although reviews of bridge inspection photos of the I–35W deck truss show that the gusset plates were bowed in the years prior to the collapse, there were no other obvious signs of deficiency. A gusset plate may be bowed due to fit up issues experienced during assembly in fabrication or in the final erection in the field. Gusset plates may also be deformed by pack rust, that is, corrosion that occurs in crevices and confined areas. Since the products of oxidation of steel have a greater volume than the base steel, it is not uncommon for bridge inspectors to see a bowing of built-up members due to rust occurring between the plates comprising the member. Previously, bowing in gusset plates was not presumed to be indicative of buckling due to the assumption that large factors of safety were used in their design. It was presumed that this would be the last element to fail.

The use of deck trusses for superstructures is typically limited to long span structures on US highways. As such, there are only a small number of these bridges on the inventory of any particular state. The typical inventory of bridges on the highway system of any state in the United States is in the thousands. In comparison, the Minnesota Department of Transportation (MnDOT) had to inspect 25 structures with deck trusses. Elsewhere in the nation, as examples, the KDOT inspected 7, while the New York State Department of Transportation reported inspecting 50 such structures.

No state reported finding evidence of buckling in gusset plates in otherwise relatively pristine condition, as on the I-35W Bridge (though MnDOT had thought this to be the case, initially, on the DeSoto Bridge in St. Cloud). What was not uncommon was finding a loss of section in gusset plates due to corrosion. With hindsight, this was to be expected. The top face of the bottom chord of a deck truss provides a location for water and any roadway salts to set adjacent to a gusset plate and to drive corrosion. This loss of section is typically obscured from visual observation by debris and the products of steel corrosion. Quantifying the section loss requires inspectors to clean the area and use ultrasonic thickness gages (Figure 1.20).

As a result of the deck truss inspections, very few bridges were permanently closed and replaced. Even in Minnesota, where an abundance of caution would be expected, only the DeSoto Bridge was replaced. At that bridge, bowing was observed in the gusset plates. Further investigation showed this to be the result of fit up issues during fabrication and erection. The capacity of the gusset plates was determined by consulting engineers to be intact; however, the previously scheduled replacement of the bridge was still accelerated. Another Minnesota bridge, the Highway 43 Bridge in Winona, was closed and reopened to traffic after repairs. Of the other two states previously mentioned for comparison, Kansas had no total closures and New York reported only one closure and replacement.

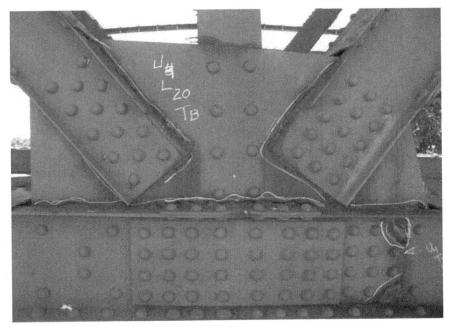

Figure 1.20 *Typical Deterioration in Gusset Plate on Truss Bridge.*

The more common outcome was the programming of repair projects for corroded and compromised gusset plates. It became a function of substantial maintenance work to repair these bridges. Repair plans and procedures were developed similar to the repair shown in Figure 1.21.

As a result of finding gusset plates compromised by corrosion throughout the nation during the inspections after the I-35W bridge collapse, bridge inspection practices were updated to focus increased attention on gusset plates in the future. In the 2002 FHWA Bridge Inspector's Reference Manual, gusset plates are discussed primarily as locations of concern for the initiation of fatigue cracking and resulting fracture. In the 2012 FHWA Bridge Inspector's Reference Manual, gusset plates are given their own section (10.8) for discussion. Fatigue issues are still discussed, but methods to determine section loss from corrosion are reviewed in detail.

However, increased inspection efforts alone would not have identified that a particular structural element had been undersized in design. That would require including the element of concern into a load rating analysis. Load rating is the analysis of a structure to determine its available live load capacity. The analysis is based on the current condition of the elements of the bridge, which may be detrimentally affected by loss of section of the

Figure 1.21 *Typical Gusset Plate Repair.*

individual elements due to either deterioration or to a traffic crash. Load raters have traditionally assumed that connections have been designed to carry the capacity of the joined structural elements, and further, that those elements were more vulnerable to deterioration from corrosion than were the connections.

On January 15, 2008, Technical Advisory 5140.29 recommended supplementing the load rating procedure for new and in-service steel truss bridges with a check of the capacity of gusset plates on the structure. Formal recommended procedures and examples for the evaluation of gusset plates were provided in the February 2009 FHWA publication, Load Rating Guidance and Examples for Bolted and Riveted Gusset Plates in Truss Bridges.

1.5.3 Cost of Failure

The most profound cost of the collapse of the I-35W Bridge was the 13 lives lost and the injury to 145 people; however, the collapse also resulted in large economic costs and impacts. With the unexpected loss of the bridge, drivers who had previously used the bridge for direct access to and from downtown Minneapolis and the University of Minnesota had to find other routes.

An initial analysis by MnDOT calculated a cost to bridge users of $400,000 per day that the bridge was not in service. This was the cost that resulted from increased travel times as automobile and heavy commercial vehicles used available detours. Of the $400,000, $153,000 was determined to be the cost to commercial traffic. Increased commercial costs are borne by businesses as increased production costs and can result in reduced economic activity by those businesses. Economic modeling by a consultant retained by MnDOT calculated the reduction in the state's economic output as $113,000 per day the bridge was not in service. The I-35W Bridge was replaced by the fall of 2008, just slightly over a year after the collapse, under an expedited design-build project. Even so, the cost to the Minnesota economy was calculated as a loss of $60 million.

The primary obligation of an engineer is to preserve public safety; however, roads and bridges are also economic investments that provide economic benefits from reduced mobility costs. The maintenance and operations of these facilities require the expenditure of economic resources. The costs incurred as a result of failures illustrate the pragmatic value and economic worth of constantly improving bridge maintenance and management practices.

1.6 MANAGING THE BRIDGE PRESERVATION PROCESS

Prior to the proliferation of automobiles, bridges on public roads faced the relatively light loads of pedestrians and horse drawn carts. Though there were notable exceptions, most bridge failures, as illustrated by the examples from the early 1900s, did not result in injuries to people. With the need to carry automobile traffic, and the use of steel and reinforced concrete, bridges on public roads became larger and more substantial. As the first bridges built for such traffic reached several decades of service, the steel and reinforced concrete members suffered from deterioration. The reaction to the failures of these bridges, with the ensuing injuries and loss of life, was the establishment of a national standard requiring regular bridge inspections and evaluations.

The activity of managing and scheduling bridge inspections and evaluations, recording and handling bridge data, and making maintenance recommendations became known as *bridge management*. By 1980, bridge owners had several cycles of bridge inspection under the NBIS, as well as any inspection data retained on their inventory from prior decades. These data allowed owners to make observations on how different bridge types and components – fashioned from different materials – resisted deterioration and challenges from the environment and from traffic. The standardized NBIS format created a nationwide data set for use. In the 1980s North Carolina and a number of states began to explore the use of computerized bridge management systems (BMS) to manage these data [18]. The FHWA sponsored the creation of two BMS systems, BRIDGIT and PONTIS. Development of the PONTIS system has continued and it is currently managed by AASHTO and has been renamed BrM (in reference to bridge management).

The development of PONTIS led to changes in bridge inspection practices. To model the progression of deterioration effectively required more refined reporting and recording of inspection findings. Bridge owners who implement PONTIS conduct bridge inspections at the element level. For example, in lieu of a single numeric value reporting to the condition state of the deck, the area of the deck in each of five defined states of deterioration is recorded. Element level bridge inspection is discussed in Chapter 3.

The development of BMS software showed that the data from bridge inspections could be used as the basis for analysis for decisions at the network level. Including deterioration models in the software allowed users to predict future needs for the network of bridges, and to model how different scenarios of maintenance and replacement would affect the condition of the network. Currently, AASHTO defines a bridge management system as, "a system designed to optimize the use of available resources for the inspection, maintenance, rehabilitation, and replacement of bridges" [18].

In the past few decades, the scope of bridge management has grown from handling bridge inspection data, to using that data to project the effect of differing funding scenarios on the health of network of bridges. To the primary goal of protecting the safety of the traveling public, has been added the goal of maximizing the effect of maintenance funds on protecting the investment in bridges.

As bridge management practices further develop, those involved in the future will need to be concerned with not simply minimizing the cost to maintain the investment in bridges but with calculating the return on the investment in bridges. As illustrated in the discussion of the I-35W Bridge collapse, the failure of bridges can impact the economy of the area served. The different bridge maintenance funding scenarios modeled in current BMS software result in different levels of bridge functionality across the network. Maintaining bridges in service, but with load restrictions on them results in restrictions on routes which heavy commercial freight may use. Maintaining functionally obsolete bridges in service may result in reduction in traffic capacity on some routes. Justifying the investment in infrastructure such as public roads and bridges is made easier when the benefits are quantified.

1.7 SCOPE AND PURPOSE OF THE TEXT

The bridge preservation process consists of three activities: inspection, evaluation, and maintenance. This text is written to those involved in scoping, scheduling, and preparing plans for bridge maintenance work. Scoping bridge maintenance work requires understanding the process of inspection and evaluation to understand the contexts of reports on bridge condition. Scheduling maintenance actions requires knowledge of typical service lives, and deterioration rates, for bridge components. Such rates are functions of the materials used, type of construction, environmental conditions, and usage characteristics. The design and plan preparation for bridge maintenance work requires familiarity with the particular maintenance actions available.

In the following chapters the reader will be introduced to basic bridge elements and materials. The typical modes of deterioration experienced by those elements will be discussed. The process of bridge inspection will be examined, with the aim of describing what information is typically available to the maintenance engineer and what may be obtained. Activities for preventative maintenance will be discussed. The design procedures for various substantial maintenance activities will be presented and examples of such provided. Bridge life cycle costing will be discussed to illustrate a rational basis for allocating resources to maintain bridge operations. The text will conclude with current practices in the management of bridge information.

REFERENCES

[1] FHWA. 2013 Status of the nation's highways, bridges, and transit: conditions & performance. Washington, DC: United States Department of Transportation; 2013.

[2] FHWA. Count, Area, Length, and ADT of Bridges by Functional Classification. FHWA Bridges and Structures [Online], Available from: https://www.fhwa.dot.gov/bridge/fc.cfm; 2014

[3] Wardhana K, Hadipriono FC. Analysis of Recent Bridge Failures in the United States. J Perform Constr Fac 2003;17:144–50. American Society of Civil Engineers.

[4] FHWA. Bridge preservation guide. Washington, DC: Federal Highway Administration; 2011. FHWA-HIF-11042.

[5] FHWA. Structure Type by Year Built Count of Bridges 2011. FHWA Bridges and Structures. [Online] June 28, 2014. Available from: https://www.fhwa.dot.gov/bridge/nbi/yrblt11.cfm

[6] FHWA. Bridge Inspectors Reference Manual. Washington, DC: United State Department of Transportation; 2012.

[7] Transportation for America. The Fix We're In For: The State of Our Nation's Bridges 2013. t4america.org. [Online]; 2013.

[8] FHWA. Comprehensive truck size and weight study. Washington, DC: United States Department of Transportation; 2000.

[9] FHWA. Recording and coding guide for the structural inventory and appraisal of the nation's bridges. Washington, DC: United States Department of Transportation; 1995, 2000. FHWA-PD-96-001.

[10] American Association of State Highway and Transportation Officials. LRFD Bridge Specifications. 7th ed. Washington, DC: AASHTO; 2014.

[11] American Association of State Highway Officials. A Policy on Design Standards. Washington, DC: AASHO; 1941.

[12] Schirmer S, Wilson T. Milestones: A history of the Kansas Highway Commission & the Department of Transportation. Topeka, KS: Kansas Department of Transportation; 1986.

[13] State Highway Commission of Kansas. Biennial Report of the Kansas Highway Commission. Topeka, vol. 1. KS: Kansas State Printing Press; 1919.

[14] NTSB. Highway accident report – collapse of US 35 Highway Bridge Point Pleasant West Virginia December 15, 1967. Washington, DC: United States Department of Transportation; 1968.

[15] NTSB. Highway accident report – collapse of a Suspended Span on Interstate Route 95 Highway Bridge over the Mianus River Greenwich, Connecticut June 28, 1983. Washington, DC: United States Department of Transportation; 1984. NTSB/HAR-84-03.

[16] FHWA. Underwater bridge inspection manual. Washington, DC: United State Department of Transportation; 2010. FHWA-NHI-10-027.

[17] NTSB. Collapse of I-35W Highway Bridge, Minneapolis, Minnesota, August 1, 2007. Washington, DC: US Department of Transportation; 2008. NTSB/HAR-08/03.

[18] American Association of State Highway and Transportation Officials. The manual for bridge evaluation. 2nd ed. Washington, DC: AASHTO; 2011. MBE-2.

CHAPTER 2

Bridge Elements and Materials

Overview

Information for a basic understanding of bridge structures necessary to program bridge maintenance and preservation work is presented in this chapter. The classification of bridges by roadways, structure type, and material is discussed. The basic elements of buried and span bridge structures are described. A basic overview of bridge mechanics is presented. The materials commonly used in bridge construction, and their modes of deterioration are discussed.

2.1 CLASSIFICATION OF BRIDGE STRUCTURES

A highway bridge by definition is a structure that carries a highway over an obstruction. Structures remain functional as long as their load–carrying capacity exceeds the demands required of them. For a highway bridge, the volume and characteristics of the traffic being carried on the highway determines the demands on the structure at any particular bridge site. Modern highways carry mostly automobile traffic and can be categorized by the restrictions in access onto and from the facility, traffic volumes, and speeds. In the United States, the American Association of State Highway and Transportation Officials (AASHTO) document, A Policy on the Geometric Design of Highways and Streets (commonly referred to as the Green Book) classifies public road facilities into the following categories:

- Interstate;
- Freeways and expressways;
- Principal arterials;
- Minor arterials;
- Major and minor collectors; and
- Local roads.

Each of these functional classifications implies criteria for the geometry of the facility including: roadway width, shoulder types, clear zones, allowable grades, and superelevations, and level of service criteria such as the allowable frequency of overtopping of the roadway from flooding. Highways classified in the higher categories of Interstate and Freeways and Expressways may be expected to carry larger volumes of traffic with more frequent and heavier trucks than roadways in the lower categories. A highway bridge must satisfy the demand of carrying the characteristic vehicular loads of a highway facility of a particular width and geometry while providing a particular clearance for any roadways or hydraulic features, which may lie beneath.

The capacity of a highway bridge structure is limited by the structural type and materials from which it is composed. The predominant characteristic of a bridge structure is the span, the distance between the vertical support over which the load must be supported. The length of a bridge span not only determines what types of structures are feasible to directly support the highway facility but also determines the magnitude of loads that must be borne by the elements supporting the span structural elements.

Inventorying a bridge structure by the functional classification of the roadway carried and by the structure type and material of its longest span allows bridge owners and operators to identify maintenance and inspection issues that may be expected for the particular bridge. For example, a steel truss bridge on an Interstate route will have to be inspected for signs of fatigue in the tension members of the truss, fatigue which may be exacerbated by heavy truck traffic. A concrete slab bridge on the same route would not be subject to the same fatigue issues.

2.2 BURIED STRUCTURES

Shorter span structures are more common on public roads in the United States than are longer span structures with the most common structures not even classified as bridges, at least by the Federal Highway Administration (FHWA) definition of being a structure with a minimum total opening of 20 ft. In the initial National Bridge Inspection Standards (NBIS) bridge inspection standards (Bridge Inspector's Training Manual 70) from 1970, all such structures were defined as culverts. However, current standards and practices recognize that culverts are not completely defined by size. A culvert may be characterized as a penetration through the berm supporting the roadway to allow conveyance through. Typically this is conveyance of water from drainage or from small streams, although it may also be for the passage of animals, pedestrians, or single lanes of vehicles. Culverts are not sized to allow the peak volume of water to pass through in free, open channel flow, but instead in submerged condition. Culverts are typically buried in the berm of the roadway with fill over the top. They lack deep foundations and are often formed from closed shapes such as pipes or box sections.

Some culverts are bridge sized. Almost 135,000 of the bridges on the 2013 National Bridge Inventory (NBI), over 22% of the total, were classified as culverts. However, the vast majority of smaller structures are culverts, and there are many more of these smaller structures on public roads. Since structures with a maximum total opening of 20 ft. or less are not required to be inspected by the FHWA, a count of such structures on all public roads in the United States is not available. But counts from some states, which have inventoried these structures, is illustrative of the scale. In 2004 in Utah, there were 47,000 culverts as compared the 2,799 bridge structures they reported on that year's NBI [1].

Given the sheer number of culvert structures, they require significant attention by anyone involved in preserving and maintaining highway bridges. The feature which most characterizes this structure for the focus of maintenance work is that they are embedded, or buried, in the road berm.

A typical reinforced concrete box culvert is shown in Figure 2.1. The primary components for a buried structure such as this are the conduit for conveyance through the roadway berm and the retaining structures at each end of the conduit. The body of the conduit may consist of one cell or several adjacent cells. In typical practice, for highway loading and with contemporary construction materials, cells are rarely wider than 20 ft. and usually no more than 14 ft. The import of this is that the top element of the conduit (for a box culvert this is the ceiling slab) is relatively lightly loaded compared to structures with longer spans and may have considerable capacity to distribute the effects of loading perpendicular to the span. This is true even for such structures, which although still buried or embedded in

Figure 2.1 *Typical Single Cell Culvert Structure.*

Figure 2.2 *Multiple Cell At-Grade Box Bridge.*

fill, may have a total opening bigger than 20 ft. or may have the top of the structure exposed directly to traffic as shown in Figure 2.2.

Although the structures are the most numerous type on the inventory of most bridge owners and operators, they are thankfully robust and cheaper to construct and maintain per unit of footprint than are longer span structures.

The box shape, constructed with reinforced concrete, is common for contemporary buried structures. It may or may not be constructed with a fillet between the horizontal and vertical elements to provide rigidity. Pipes are another common structure and may be formed from reinforced concrete, corrugated steel or aluminum, fiberglass, or plastic (Figures 2.3 and 2.4).

Figure 2.3 *Fillet on Reinforced Concrete Box Culvert.*

Figure 2.4 *Multiple Corrugate Metal Pipes.*

Less common are arch shapes, constructed from reinforced concrete or corrugated steel. This structure requires the construction of a footing for each wall, but the open bottom allows for a more natural stream passage, which may be of use to meet environmental permitting requirements (Figure 2.5).

The foundations of buried structures are typically founded in the soil mantle and do not extend to bedrock material as is more common is longer spanned structures. This contributes to lower cost, but may present issues

Figure 2.5 *Inside of Reinforced Concrete Arch.*

Figure 2.6 *Reinforced Concrete Box With Concrete Apron.*

due to undermining of the support mantle due to the flow of water at either the entrance or the exit of the conduit. To protect the supporting soil
mantle, a concrete apron may be placed at either the entrance or exit end of
the conduit, depending on the need. A similar effect may be achieved with
an apron of large rock (Figure 2.6).

If there is a need to prevent degradation of the streambed of the channel
approaching the structure, a soil saver may be constructed. This is a small
wall to retain the fill beneath the channel and maintain its elevation. This
reduces the slope of the channel approaching the conduit and reduces the
velocity of the water (Figure 2.7).

Figure 2.7 *Inlet End of Reinforced Concrete Box With Soil Saver.*

Figure 2.8 *Timber Headwall on Corrugated Metal Pipe.*

For pipe structures, the structure retaining and protecting the fill around the conduit is a headwall. The headwall may also serve the purpose of directing the flow into the pipe, resulting in a more orderly flow and minimizing turbulence. This improves the capacity of the conduit for conveyance of water (Figure 2.8).

Box structures, arches, and large–opening pipes will have retaining walls at each end to retain fill. Typically these are constructed of reinforced concrete; however, particularly for structures, which have been repaired on an expedient basis, they may be formed in a number of manners. They may be formed by piles and lagging or even by blocks massive enough to resist earth pressure by weight alone (Figure 2.9).

The angle at which these walls are oriented with respect to the conduit may serve to direct flow or may serve to provide clearance for ditches or other adjacent features. As such, the angle may vary from parallel to the conduit to parallel to the direction of the roadway (Figure 2.10).

Figure 2.9 *Concrete Wingwall and Temporary Wingwall.*

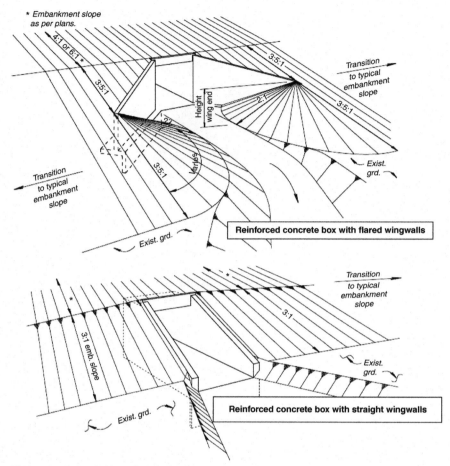

Figure 2.10 *Straight and Flared RCB Wings. (From KDOT Bridge Design Manual [2]).*

2.3 ELEMENTS OF SPAN BRIDGE STRUCTURES

Although culverts constitute a large portion of the inventory of structures on most highway systems, when visualizing a bridge most people think of a span structure. Rather than being characterized as a penetration through the road berm, for a span structure the berm clearly stops where the span starts.

The typical elements of a span bridge for a highway are shown in Figure 2.11. The elements are divided into those that are a part of the substructure and to those that are a part of the superstructure of the bridge. The substructure consists of all those elements that support the superstructure. The elements supporting the ends of the superstructure and retaining the

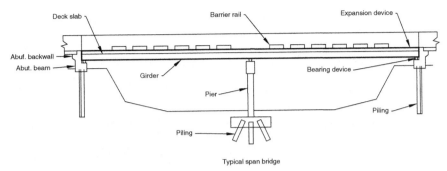

Typical span bridge

Figure 2.11 *Typical Parts of a Span Highway Bridge.*

embankment are abutments. The intermediate supports are piers. The superstructure consists of those elements that carry the roadway from bearing point to bearing point on the substructure elements.

2.3.1 Substructure Elements

2.3.1.1 Abutments

On a span bridge, the wearing surface of the roadway is the top surface of the deck element in the superstructure. Differential settlement between the bridge and the embankment berm under the approach roadway can result in a bump in the road at the start of the bridge. Depending on the magnitude of the settlement, this bump may be an annoyance to drivers or can pose a danger to the safe operation of motor vehicles. Proper construction and maintenance of abutments are critical to prevent this (Figure 2.12).

On modern highway bridges, the abutment is typically constructed from reinforced concrete. The portion of the abutment on which the end of the superstructure bears is referred to as the abutment beam or seat. The portion which retains the embankment fill directly behind the superstructure

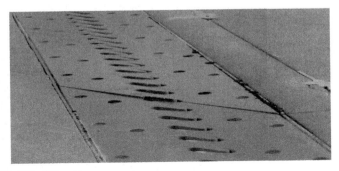

Figure 2.12 *Tilted Finger.*

Figure 2.13 *Bearing Seat on Abutment Backwall.*

is the abutment backwall. In the configuration shown in Figure 2.13, the superstructure may move horizontally independently (up to the gap distance) of the abutment. This allows for thermal expansion and contraction of the superstructure, which depends on the length and material of the superstructure. For long steel structures, the range may be well over 4 in.

The separation between the deck and the abutment backwall allows for water and debris to fall onto the abutment-bearing seat and can be an ongoing source of deterioration for the abutment and the superstructure element. For span structures with shorter spans and smaller thermal movements (less than 2 in.) the abutment may be formed integrally with the superstructure as shown in Figure 2.14. The caveat is that the foundation

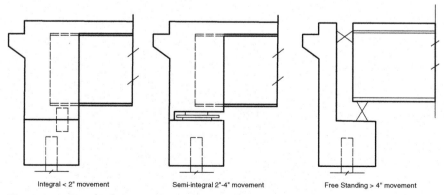

Integral < 2" movement   Semi-integral 2"-4" movement   Free Standing > 4" movement

Figure 2.14 *Types of Abutment Continuity With Superstructure.*

element supporting the abutment must be capable of sustaining the lateral movement required without damage.

For movements beyond what the foundation elements can withstand without damage, the abutment backwall may be formed integrally with the superstructure and configured to move independently of the abutment beam. This style of abutment is referred to as semi-integral.

Any portion of the abutment cast initially integrally with the superstructure may be referred to as part of the superstructure, as with the entirety of the integral abutment in the Figure 2.14. However, an integral abutment may be built by placing the abutment beam first, then setting the superstructure on top and casting the portion above the beam integral with the rest of the superstructure. Although, as shown in Figure 2.14, the entire cross-section of the abutment must move with the superstructure, the beam would be considered as substructure.

An abutment may also be classified by its relation to the embankment fill (Figure 2.15). An abutment which sits atop and in the embankment fill is referred to as either a stub, perched, or shelf-type abutment (in Figure 2.16A). An abutment in which a wall bears the weight of the superstructure down to the foundation and retains the full height of the embankment fill is a full height, cantilever, or closed abutment. Abutments with structural elements separating the beam near the top of the embankment and a footing under the embankment are spill-through or open abutments. Although the

Figure 2.15 *Integral Abutment With Straight Wings on Girder Bridge.*

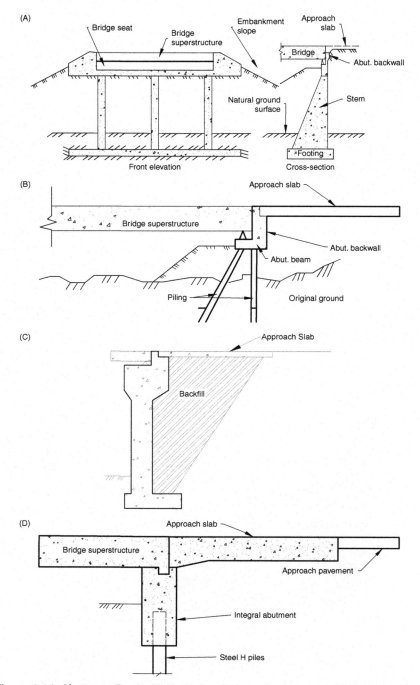

Figure 2.16 *Abutment Types.* (A) Typical spill through abutment. (B) Typical stub or perched abutment. (C) Typical full-height closed or tall abutment. (D) Typical integral abutment. *(Adapted from NCHRP Synthesis 234 [3]).*

FHWA Bridge Inspectors Manual classifies integral abutments as a separate type from the previously mentioned three classifications, in practice integral abutments are almost always a subset of stub abutments. It is rare that the thermal movement to be accommodated is small enough that a wall or similar reinforced concrete element could do so without being damaged.

The embankment berm fill to the outside of the road is retained by abutment wings. For small differences in elevation between the top of the abutment and the berm under the bridge superstructure, these wings may be straight, that is, in line with the abutment backwall. For large differences in elevation, typically due to deeper superstructure elements, the wings may be turned back, that is, parallel to the roadway. Abutment wings may also be flared, or oriented at any angle in between, as required by the conditions at the bridge site or in the roadside environment.

Although abutments are typically founded on deep foundation elements, discussed in Section 2.3.1.3, to provide the vertical stability that is paramount to minimizing differential settlement, in recent years some highway bridge abutments have been built atop embankment fill that has been mechanically stabilized. In this layout, a stub abutment is constructed atop a mass of stabilized fill, which is over the *in situ* material. The concept rests on the premise that any difference in settlement between the bridge and the approaching roadway will be spread across a long enough section of roadway above the fill to minimize the impact on vehicle operation (Figure 2.17).

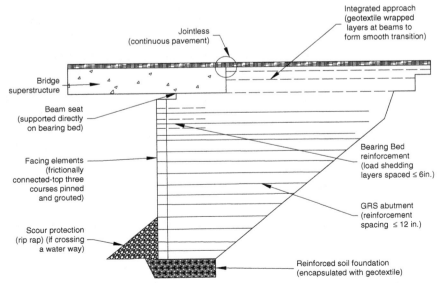

Figure 2.17 *Typical Plan of GRS Abutment. (Adapted from FHWA-HRT-11-027 [4]).*

2.3.1.2 Piers

The substructure elements supporting the spans between the embankment berms are piers. Although the term pier is commonly used to refer to all such elements, the FHWA Bridge Inspector's Reference Manual does make a distinction between piers and pier bents (or simply, bents). As defined in the manual, a pier has a common foundation element, while a pier bent has multiple foundation elements (Figure 2.18).

The FHWA Manual defines four common pier configurations: wall (solid), columns, columns with web walls, and cantilever (hammerhead). By far, the most commonly used material in construction of piers is reinforced concrete. Stone and masonry wall piers are not uncommon on smaller span

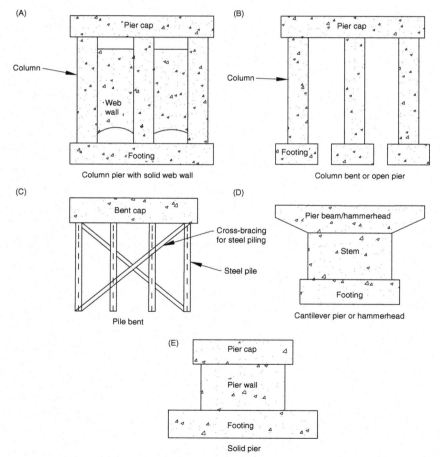

Figure 2.18 *Pier and Bent Types. (Adapted from FHWA Bridge Maintenance Manual [5]).*

Figure 2.19 *Steel Pier Beam on Straddle Bent.*

bridges constructed in the first decades of the last century or earlier. In contemporary designs, where highway geometrics result in constraints on pier beam depths, steel may be used to allow for shallower members than may be formed from reinforced concrete (Figure 2.19).

Pier bents come in two configurations, column bents and pile bents. The difference being that the pile in a pile bent typically acts as a foundation element. Reinforced concrete is the most common material for pier-bent construction, but pile-bent construction in particular may vary with the use of steel, prestressed concrete, and timber piling.

2.3.1.3 Foundations

To prevent settlement, most modern highway bridges are founded on elements that carry the bridge's loading and weight well below the soil mantle and into either bedrock or undisturbed geological material, which will bear the load without settling. This is a primary difference between most span structures and buried structures. Foundations that extend below the mantle are called deep foundations.

If bedrock material is near the ground surface, abutments and piers may be founded on spread footings. These spread the load of the structure over direct bearing on the rock member. Limiting strengths are typically on the order of 2000–6000 psf.

If rock is unavailable near the surface, pilings may be used. Pilings are typically steel (Figure 2.20), prestressed concrete (Figure 2.21), or timber members, driven through the soil mantle into underlying geological

Figure 2.20 *Steel H-Piles.*

members where they reach capacity by either point bearing on rock, by friction against the member driven into, or by a combination of both.

Another common deep foundation element is the drilled shaft or caisson (Figure 2.22). Individual shafts vary from 2 ft. to 10 ft. in diameter. Shafts are formed by auguring out geological material down into the supporting layer. Reinforced concrete is then cast in place. Either a permanent or temporary steel casing may be used to prevent collapse of surrounding geologic material prior to concrete placement and/or the infiltration of water into concrete as it is placed. Similar to piling, the shaft may develop load-carrying capacity by bearing on bedrock, by friction against the rock around the perimeter near the base of the shaft, or a combination of both.

Figure 2.21 *Concrete Piles.*

Figure 2.22 *Drilled Shafts.*

On contemporary highway bridge structures, smaller diameter reinforced concrete shafts may also be found. These will have been placed by either augercast (typically 12–24 in., but can be up to 42 in. in diameter) or micropile (3–12 in. in diameter) methods. For augercast piles, once the auger reaches its desired depth grout or concrete is pumped through the stem of the auger as it is pulled from the hole. Reinforcing steel is placed afterward. For micropiles, a casing is drilled to a desired depth, then reinforcing is placed to that depth or farther as required to develop capacity, and then the casing is filled with concrete or grout (Figure 2.23).

2.3.2 Superstructure Types and Elements

The superstructure consists of the elements carrying traffic loads across the spans to bearing on the abutments and piers, and the appurtenances attached to these elements. In the most minimal configuration, this consists of a structural slab and railing. In the common case of a beam or girder type of superstructure, this consists of the railing, deck, beams or girders, expansion joints, and bearings (Figure 2.24).

2.3.2.1 Railing

All modern span bridges carrying highway traffic have railings to keep traffic that attempts to depart from the roadway from falling over the side of the bridge. Unlike guard fence found on the roadside environment and which may be used on roadways over buried structures, bridge rails are designed to be rigid and not to deflect upon impact. Instead impacting vehicles are redirected and rails must be capable of resisting fairly large loads.

Figure 2.23 *Installation of Micropile.*

There are various shapes and configurations of bridge rails and the common materials are reinforced concrete, steel, and aluminum (Figures 2.25 and 2.26). Contemporary bridge rail configurations are crash tested for vehicular impacts and are of significantly more substantial construction than rails from only a few decades ago (Figure 2.27). Newer rails may require

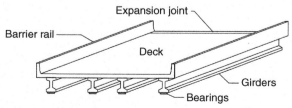

Figure 2.24 *Superstructure Components on Beam Bridge.*

Figure 2.25 *Corral Rail.*

attention due to damage from impacts, but older rails may also require attention due to deterioration abetted by their lighter construction.

2.3.2.2 Decks

For most span bridge configurations, the element that directly supports traffic is the deck. Most modern highway bridge decks are formed from reinforced concrete and are typically 6–10 in. in thickness. The wearing surface exposed to traffic may be the top face of the deck slab or there may be an

Figure 2.26 *Aluminum Rail.*

Figure 2.27 *Damaged Type A Rail.*

overlaying layer of material. Typical overlay materials vary from a thin layer of polymer with aggregate broadcast across the surface to provide for traction, to 1–2 in. of high density, low permeability concrete, to a few inches of bituminous material (Figure 2.28).

Exposed to wear from dynamic traffic loads and subject to exposure from the weather and from deicing treatments, the deterioration of bridge decks is often the primary focus of maintenance efforts for bridge structures. Overlays applied during the life of a bridge are a common maintenance

Figure 2.28 *Wearing Surface of Deck With Polymer Overlay.*

Figure 2.29 *Timber Deck.*

action used to deal with deterioration in the deck material and to restore a smooth riding surface for traffic.

Although reinforced concrete decks are by far the most common type of construction, other deck materials may be encountered in an inventory of highway bridges. On bridges carrying a very low volume of traffic, timber decks may be found (Figure 2.29). On long span bridges, where the weight of the deck may be a significant concern in the structure design, steel grates have been used (Figure 2.30). On newer structures, deck constructed from

Figure 2.30 *Filled Steel Grate Deck.*

Figure 2.31 *Wearing Surface of Extruded FRP Deck Panels.*

fiber resin polymers (FRP) have been used in an attempt to address corrosion and weight; however, to date there is not a long history to judge FRP performance (Figure 2.31).

An item of particular interest to those who deal with bridge maintenance is deck drainage (Figure 2.32). Provisions must be made to convey precipitation away from driving lanes and off the deck. If a bridge is relatively short, curb or barrier rail may direct drainage down the curb line of the deck and off of the approach at the downhill end of the bridge. Alternately, open railing may allow runoff over the side of the deck. However, if a bridge is long enough and open rail cannot be used in order to prevent the spread from runoff into driving lanes; drains through the deck must be provided. Deck drains and associated piping come in a myriad of forms. Common to most drains is a steel grate over the top surface to prevent wheels from dropping into the drain pan. Maintaining deck drains is a key issue in preventative maintenance.

Figure 2.32 *Deck Drain.*

2.3.2.3 Expansion Joints

Even in the most mild of climates, highway bridges are subjected to temperature changes in the environment that result in measurable expansion and contraction of spans over the course of a season. These thermal movements are significant enough to necessitate accommodation by the structure. For every 10°F increase in ambient temperature, a steel superstructure will expand 0.078 in. per 100 ft. of span length, and will contract an equal amount during a 10° drop. A 200-ft. long steel bridge is over 1.5 in. longer when the temperature is 100°F than on a 0° day. Bridges with concrete superstructures will also expand and contract. The coefficient of thermal expansion of concrete is similar to that of steel, but concrete bridge members typically lose heat more slowly compared to their less-massive steel counterparts. As a result a concrete bridge will expand and contract less over a given change in ambient temperature than a steel bridge.

In bridge decks, accommodation for thermal movement is often provided by expansion joints. There are a number of configurations of joints, which vary from a simple open gap to assemblies of multiple elements including springs. A primary consideration in the design of all of expansion joints is that the maximum transversable gap across the driving lane on a highway bridge is 4 in. Gaps larger than this pose a hazard to the wheels of smaller vehicles, such as motorcycles.

The FHWA Bridge Inspector's Reference Manual sorts expansion joints into six categories:
- Strip seal expansion joints;
- Pourable joint seals;
- Compression joint seals;
- Assembly joints with seal (modular);
- Open expansion joints; and
- Assembly joints without seals (finger plate and sliding joints).

All deck joints are subjects of particular interest in bridge maintenance in that their failure effects not only the ride on the wearing surface, but the water and debris that flows through failed seals onto elements below may result in significant deterioration.

2.3.2.4 Bearings

Except for configurations where the superstructure is integral with the substructure the framing of the superstructure sits atop bearing devices, which sits atop the substructure elements. These devices isolate the substructure elements from the deflection and rotation of the superstructure in response to loading, and sometimes from the thermal movements of the superstructure, while providing a load path for support. A bearing which acts as a pin, allowing rotation while restraining the superstructure from moving horizontally, is a fixed bearing. Bearing devices which allow horizontal translation and rotation are expansion bearings.

Given the large loads that must pass through a bearing device, prior to 1950 almost all highway bridge bearing devices were constructed from steel, as are a large number of bearings since then. A fixed steel bearing is a bolster. In the bolster shown in Figure 2.33, the four basic components of most bridge bearings are labeled. Those components are the sole plate (attached to the superstructure), the bearing device, the masonry plate (attached to the substructure), and the anchor bolts (which holds the masonry plate in place and may be used in other bearing configurations to restrain the bearing device).

Steel bearings, which allow for expansion may be called movable bearings, in that expansion is allowed by providing for movement in the bearing device. The three basic configurations are sliding, rockers, and rollers – each of which allows for progressively more movement.

The material found progressively more commonly in bridge bearings since 1950 is elastomer. The term elastomer is derived from elastic polymer, referencing the characteristic of the material to return to its original shape

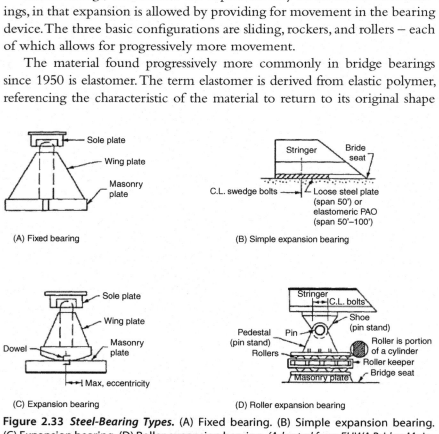

(A) Fixed bearing

(B) Simple expansion bearing

(C) Expansion bearing

(D) Roller expansion bearing

Figure 2.33 *Steel-Bearing Types.* (A) Fixed bearing. (B) Simple expansion bearing. (C) Expansion bearing. (D) Roller expansion bearing. *(Adapted from FHWA Bridge Maintenance Manual [5]).*

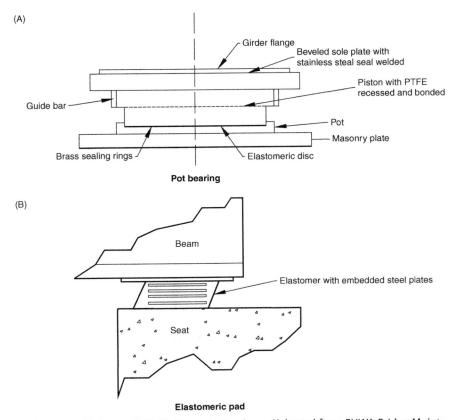

Figure 2.34 (A) Pot and (B) Elastomeric Bearings. *(Adapted from FHWA Bridge Mainte-nance Manual [5]).*

after load is removed. The most common elastomers used for bridge bearings are neoprene and natural rubber, with neoprene being the more commonly used in the United States. To resist higher stresses the elastomeric material may be reinforced with layers of steel (Figures 2.34 and 2.35). Whether used as fixed or as expansion bearings, elastomeric bearings may be thought of as deformable bearings. Rotational movement is accommodated by deformation of the elastomer under load. Translation may be accommodated solely by deformation of the elastomer for small movements, or by use of a polytetrafluoroethylene (commonly known as PTFE or Telfon™) sheet to permit sliding.

To provide for the capacity to carry high-bearing loads and for rotational capacity in multiple orientations, pot bearings and disc bearings are used. In the pot bearing, the elastomer is confined within a pot – a steel ring – with the load transmitted through a piston attached to the sole plate

Figure 2.35 *Reinforced Elastomeric Bearing.*

(Figure 2.36). Disc bearings are similar save that a hard plastic disc is used in place of the confined polymer.

There are multiple variations and permutations of bridge bearings from the previously described categories and there are several specialty bearing designs, particularly to accommodate seismic demands. A complete description of all types is beyond the scope of this text.

2.3.2.5 Span Structures

The element which does the most to define the character of the bridge is the structure of the span. Although the structure type for some bridges is chosen to reflect an aesthetic statement, typically it is the economics of construction and the familiarity of designers with particular structural designs that are the primary factors behind selection of the span type. Different span

Figure 2.36 *Pot Bearing.*

types are at their lowest cost to construct for different span lengths. Considerations that determine the construction costs for the span structure include the total amount of material required, the cost to fabricate the materials and components for the span, and the cost to erect the structure in place.

2.3.2.5.1 Structural Slab

The simplest and cleanest type of superstructure is the structural slab. The span of a structural slab is formed by a deck with any intermediate framing between it and the substructure. Since the entire width of the slab is available to provide a load path to support vehicular loads, this structure type is highly robust. They are limited in application to shorter spans, less than 40 ft. for individual simple spans and less than 70 ft. as part of a series of continuous spans (Figure 2.37). See discussion in Section 2.4 for definition of simple versus continuous spans.

Typical construction of a structural slab includes forming from cast in place reinforced concrete. The amount of falsework and formwork required to place these structures can limit their usage to locations where the area below the span may be obstructed temporarily.

2.3.2.5.2 Beam or Girder

The structural form most commonly used in highway bridges to span short- to medium-length spans (spans less than 400 ft.) are beams or girders (Figure 2.38). Both terms refer to members that act as beams in terms of their structural mechanics, that is, they are primarily loaded transversely to their long axis, which induces flexure, or bending, in the member. In the

Figure 2.37 *Single Span Reinforced Concrete Haunched Slab Bridge.*

Figure 2.38 *Steel Girder Bridge.*

nomenclature of bridge elements, girders differ from beams in that girders are typically deeper than beams. The greater depth of girders make them suitable for use on longer spans (usually over 100 ft.) but makes them less compact and makes stability of the section in buckling a concern.

Beams may be fabricated (off-site) or formed (at the bridge site) from rolled steel sections, cast in place concrete, timber, or prestressed concrete (Figure 2.39). Girders may be fabricated from plate steel, formed from cast in place concrete (typically for concrete box girder bridges) or erected from prestressed concrete sections posttensioned together at the bridge site. Beam

Figure 2.39 *Prestressed Concrete Beam Bridge.*

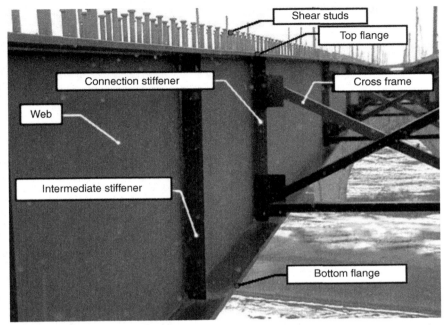

Figure 2.40 *Typical Framing Components for Steel Girder Bridge. (From KDOT Bridge Construction Manual [6]).*

and girder spans may be either simple or continuous, though simple girder spans are relatively rare (Figure 2.40).

Steel girder spans may have secondary framing consisting of stringers and floorbeams to carry the load from the deck to the girders (Figure 2.41). Almost all beam or girder bridges have diaphragms or cross-frames between beam lines

Figure 2.41 *Stringer and Floorbeam System.*

Figure 2.42 *Welded Steel Rigid Frame Bridge.*

to brace the beams against twisting and to facilitate the distribution of loading across the system of beams. On longer spans, a system of lateral (or sway) bracing may be found to stiffen the span for wind loads and for construction.

2.3.2.5.3 Rigid Frames

If the structure of the span is integral and rigidly connected with the supporting substructure elements, and the span and the substructure element are of similar stiffness, then a rigid frame is formed (Figure 2.42). For a steel structure this most commonly takes the form of plate girders fabricated to be monolithic with legs at pier locations. For concrete structures this most commonly takes the form of a structural slab cast integrally atop walls at the abutment locations.

If the superstructure is significantly stiffer than the substructure elements, then even if both elements are formed integral they are not considered a frame. The structural slab of the reinforced concrete haunched slab shown in Figure 2.43 is integral with the pier beam and columns; however,

Figure 2.43 *Reinforced Concrete Haunched Slab on Columns.*

the stiffness of the columns does not significantly affect the behavior of the slab above; therefore it is not considered a frame-type structure.

2.3.2.5.4 Truss

A truss is an assembly of members, which acts as a beam, that is, it is primarily loaded in flexure. However, the individual members are slender and are assumed to be loaded only in axial tension or compression. A truss makes efficient use of material and is economical for new construction of spans from 450 ft. to 900 ft. in length. Much shorter trusses may be encountered on many bridge inventories. However, the current costs associated with fabrication and erection of the various parts of trusses offsets the material savings for spans less than 450 ft.

There are several configurations of trusses, such as Pratt or Warren, as illustrated in Figure 2.44. In some configurations, the diagonals are in

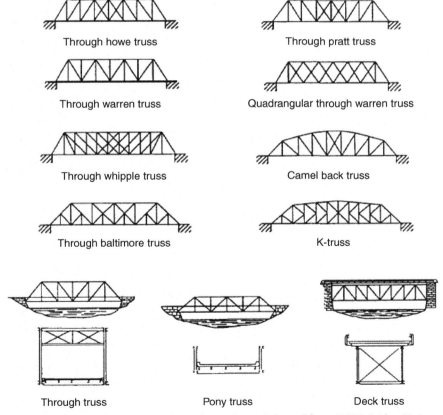

Figure 2.44 *Various Truss Bridge Configurations. (Adapted from FHWA Bridge Maintenance Manual [5]).*

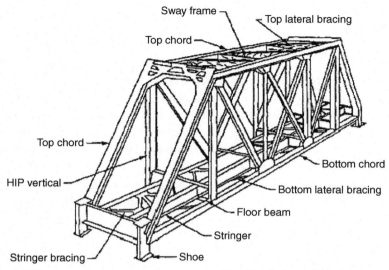

Figure 2.45 *Truss Components.* *(Adapted from FHWA Bridge Maintenance Manual [5]).*

compression and the verticals are in tension; in others it is vice-versa. Trusses may also be categorized by whether the deck is supported by secondary framing on top of the trusses (deck truss), above the bottom chord of trusses (through truss), or in between (pony truss). With either the deck or the through trusses, bracing may be placed between the trusses in the plane of the top and bottom chords. Trusses so stiffened are suitable for long spans. Without bracing across the top chord, pony trusses are suitable only for fairly short spans of 100 ft. or less (Figure 2.45).

Almost all highway bridge trusses are constructed from steel, the rare exception being some historic wood or wood and metal trusses such as found in covered bridges.

2.3.2.5.5 Arch

The arch is a classic structural form, which resolves the vertical loading of a span into compression in its rib. The arch may be constructed from stone, masonry, concrete, or steel. Stone and masonry arches may be found on inventories with older bridges as culvert-sized structures (Figure 2.46). The roadway is typically supported on earthen fill above the arch contained between walls, which close the spandrels. The spandrel is the area between the roadway and the arch.

The use of reinforced concrete or steel allows for longer arch spans with an open spandrel framed by vertical members to carry the load of the

Figure 2.46 *Stone Arch Bridge.*

deck to the arch. The use of reinforced concrete and steel also allows for the arch to be constructed above the roadway, making for through arches (Figure 2.47).

The compressive force in an arch rib must be resisted to maintain the shape and stability of the arch. In a true arch structure this is done at the footing of the arch. Alternately, with reinforced concrete or with steel construction, a horizontal tie may be used, resulting in a tied arch (Figure 2.48).

Figure 2.47 *Reinforced Concrete Arch Bridge.*

Figure 2.48 *Network Tied Arch Bridge.*

For new steel construction, a tied arch is economical in spans ranging from 450 ft. to 900 ft. True arches may be used for spans up to 1800 ft., as with the Lupu Bridge in China for example.

2.3.2.5.6 Complex Bridges

Bridges for which the main span is supported from cables made up less than 0.05% of the total number of bridges in the FHWA National Bridge Inventory of 2013. There were 96 suspension bridges, from which the span is supported from vertical cables, which attach to the main suspension cables spanning between pylons. There were also 48 cable-stayed bridges, where the supporting cables run from the girders directly back to the pylons (Figure 2.49).

Cable-supported bridges are among those types categorized by the FHWA as complex bridges. The other major category of complex bridge is the moving bridge. On moving bridges, typically a span is moved by a mechanism when required to make way for shipping on a waterway below the span. This mechanism may turn the span parallel to the waterway (swing bridge), may open opposing sections of the span upward (bascule bridge) or may move the span upward on vertical towers at each end of the span (lift bridge). Although at a count of 833 in 2013, there are considerably more moving bridges than cable-supported bridges on the NBI, they still are less than 0.25% of the total number of bridges on US public roads.

Complex bridges make up only a very small part of the total bridge inventory and require special inspection and maintenance protocols for particular elements on the bridges. Although such bridges share many of

Figure 2.49 *Christopher Bond Bridge – Cable Stayed.*

the same inspection and maintenance challenges as the majority of highway bridges, the particularities of complex bridges are not discussed in this text.

2.4 BRIDGE MECHANICS

The structure of a highway bridge must support and resist multiple substantial loads, including the weight of traffic and of the bridge, wind, the braking and acceleration of traffic, etc. Although the individual members of a bridge deform (i.e., change shape) under loads, they must remain in a static position. To maintain this static equilibrium internal forces are developed in the members of a structure, providing a load path exists from the externally applied loads to the reactions on the foundation and substructure elements (Figure 2.50).

The calculation of the magnitude of these internal forces requires the application of techniques of structural analysis, which is beyond the scope of

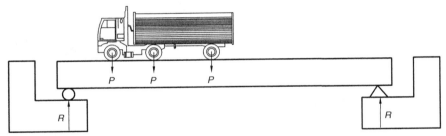

Figure 2.50 *Traffic Loading and Reactions on Span Bridge.*

this text. However, understanding the forces typically resisted by particular bridge elements can help those involved in bridge maintenance understand how the demands exerted by those forces effect the performance of those members, and how deterioration and failure may manifest.

2.4.1 Axial Forces – Tension

The simplest force state to consider is that of pure tension (Figure 2.51). A tensile force is one, which tends to elongate a member. Pure tension occurs in cables and is assumed to occur in the tension members of trusses. These members are pulled along their primary axis, are not loaded perpendicular to the primary axis, and are not rigidly connected at their ends.

Away from the point of application at the ends of the member, the tensile force is spread across the cross-sectional area of the member. The intensity of the force per unit area is the stress at that section of the member. When the tensile force is relatively evenly spread across the entire cross-section the stress is equal to the force divided by the cross-sectional area.

$$\sigma = \frac{Force}{Area} = \frac{P}{A}$$

The elongation that results from the application of a tensile force can be defined in terms of the ratio of the change in length to the original length of the member, a unitless quantity called strain.

$$\varepsilon = \frac{Change\ in\ Length}{Original\ Length} = \frac{\Delta L}{L}$$

When a tensile member is loaded and elongates, but returns to its original length after the load is removed, it is said to have behaved elastically. The maximum stress under which a material will behave elastically is its yield stress. Stressing a material past its yield stress results in permanent plastic deformation (Figure 2.52). Unloading a cable or a rod for which the entire

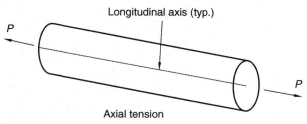

Figure 2.51 *Axial Tension.*

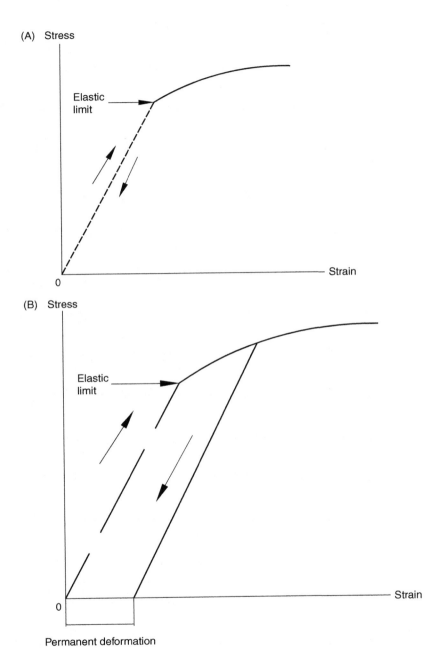

Figure 2.52 *Stress–Strain Diagram.* (A) Elastic strain. (B) Plastic strain.

cross-section has yielded would return the member to its original length plus some plastic set.

For an elastic material, the relationship between stress and strain is linear. The value of that ratio is a property specific to that material known as the modulus of elasticity.

$$E = \frac{stress}{strain} = \frac{\sigma}{\varepsilon}$$

Almost all pure tension members on highway bridges are formed from steel. Exceptions include wrought iron on historic structures and wood on old wooden trusses, such as found in covered bridges. Steel has considerable strength in tension and can undergo significant plastic strain before rupturing. Modern steels are designated by grades, which correspond to their assumed yield strength; for example, a grade 36 steel is assumed to have a yield stress of 36,000 psi. A pure tension member formed from steel, which is loaded past yield will deform prior to rupturing once the material reaches its plastic limit.

The characteristic of steel to deform prior to rupture makes it particularly useful for bridge construction. First, it provides an opportunity for inspectors to observe whether a tension member has been or is being overloaded by looking for necking, that is, a reduction in the section and thickness of a member as it is stretched plastically. Second, deformation under load once a member has started to behave plastically results in the shedding of load from the member. Taking the example of a steel tension member in truss, which has started to deform plastically will "shed" its load to adjacent members. This increases the stresses in those members, but provides a degree of redundancy for the performance of the structure (Figure 2.53).

2.4.2 Axial Forces – Compression

A force that acts to shorten a member induces compression in that member. The same concepts of stress and strain discussed in the section 2.4.1 on tension apply to compression with an important caveat: the strength and elasticity of a material may not be the same for compression as it was for tension. For example, concrete has considerable strength in compression. A compressive strength of 4000 psi is commonly assumed for the design and analysis of concrete members. However, it is weak in tension; unreinforced concrete can only resist about one-tenth of the stress in tension as it can in compression. Additionally, it is incapable of significant deformation in tension and therefore will fail relatively quickly (Figure 2.54).

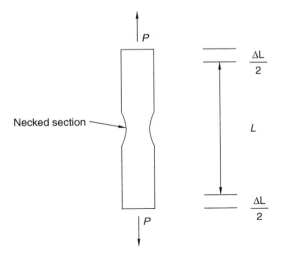

Figure 2.53 *Elongation and Necking Under Tension.*

A member in pure compression can fail by reaching yield stress across the entire cross-section of a member, as with tension. However, this is rarely the failure mode of concern. A member in tension tends to align itself concentrically with the applied force. A member in compression must maintain its alignment as the force acts to push its constituent material together. Eventually, the strain energy in the member reaches a point to where the member buckles. It deforms into a different alignment with a lower resulting strain energy. A member's resistance to buckling depends not only on the properties of the material, but to geometric properties of the member and the degree of restraint against rotation at the ends of a member.

The limiting force of a member for buckling is typically described by the Euler buckling formula.

$$P_{cr} = k \frac{(\pi^2 EI)}{L^2}$$

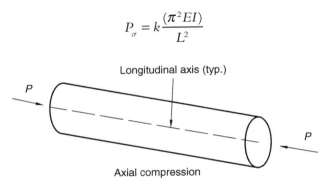

Figure 2.54 *Axial Compression.*

Where, k, effective length factor; E, modulus of elasticity of the material, I, moment of inertia of the cross-section; and L, length of the member.

The moment of inertia is a measure of the stiffness of a section and depends on the geometry of the cross-section. A member which is longer or which is less stiff (i.e., a lower moment of inertia of the section) is more slender and has a lower critical buckling load.

Compression members in a truss and the pins in a steel-bearing device are assumed pure compression members. Much more common are members, which are significantly loaded in compression, such as columns, stems, and walls at piers and abutment, and the piles and shafts in the foundations. A steel member that has buckled would twist out of alignment. A concrete member that has buckled would crack and its movement out of alignment may or may not be perceptible to a quick visual inspection (Figure 2.55).

2.4.3 Forces From Transverse Loading – Bending

Other than cables, and what is assumed for truss members, most bridge members are not loaded concentrically along their primary axis. Most have at least some loads applied transversely to the primary axis. A rigid body that is not restrained by a force will tend to rotate, unless the force(s) acting perfectly through the center of gravity of the body. The effect of a force to cause rotation or, in the case of a restrained rigid body, bending of it is called moment. A moment that causes rotation about the primary axis is commonly referred to as a torque (Figure 2.56).

In structural analysis, a member which is loaded primarily in bending (also referred to as flexure) is referred to as a beam. Hence, there may be some confusion in nomenclature when referencing bridge girders, stringers, and beams since they are all beams from the point of structural analysis.

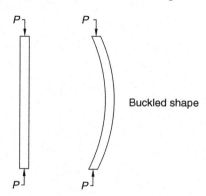

Figure 2.55 *Buckling.*

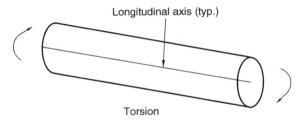

Figure 2.56 *Torsion on Beam.*

A load transversely applied to a beam, such as the self-weight of the bridge superstructure or the weight of vehicular traffic applied to a girder, results in bending the beam into a deflected shape. In this shape one side of the beam is elongated, while the opposite side is shortened. Consequently, bending results in tensile and compressive stresses across the cross-section of a beam (Figure 2.57).

The formula describing the maximum bending stress at the extreme fiber for typical beams is given below. Note that some of the assumptions used in its derivation are that deformations are relatively small as compared to the length of the beam, that plane sections in the beam remain plane and that the section remains elastic. The formula illustrates that the bending stress in a beam is inversely proportional to the stiffness of a beam for a given moment.

$$\sigma_{bending} = \frac{Mc}{I}$$

Where, M, the moment at the section; c, distance from the centroid of the section to the extreme fiber; and I, the moment of inertia of the section.

For a given amount of material, a deeper beam is more stiff and more of the cross-sectional area is farther away from the centroid, resulting in a stiffer beam. This is why beam sections are commonly shaped like a capital I, with most of the material out at the flanges. It is also why depth is so critical for members used to frame the span such as trusses, girders, and arches.

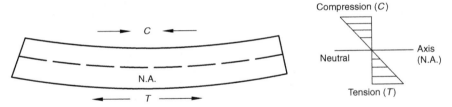

Figure 2.57 *Bending Moment in a Beam.*

Calculations are required to determine the exact magnitude of stress in response to a given loading on a structure, but one can intuitively determine the locations of compression and tension along the length of a beam by imagining its deflected shape for given location of loading and restraint. The compression side of a beam is subject to the same concerns for buckling as is a member axially loaded in compression. Similar tension concerns exist on the tensile side of a beam.

Beams which run only from one bearing point to the other are simple beams. If uniformly loaded, the location of maximum stress and deflection will be at the middle of the span. Simple beams will also be in compression on the top and in tension on the bottom for their entire length. A beam which runs over several supports is considered continuous, such as a single girder spanning from abutment to abutment across one or more piers. Over the points of interior support, the tension and compression sides of the beam will switch with the curvature of the deflected shape. The advantage of using continuous beams in bridges is the reduction of moment in the span for a given loading versus the simple span configuration, and the redundancy provided by continuity (Figure 2.58).

2.4.4 Transverse Loadings – Shear

Tensile and compressive stresses are extensional stresses. They are produced as a tiny element of the member is either lengthened or shortened. The

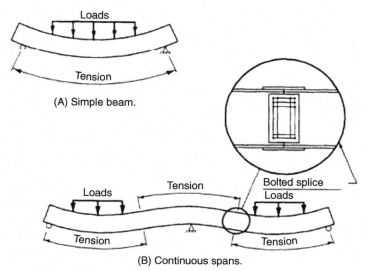

Figure 2.58 (A) Simple and (B) Continuous Spans. *(From FHWA Bridge Maintenance Manual [5]).*

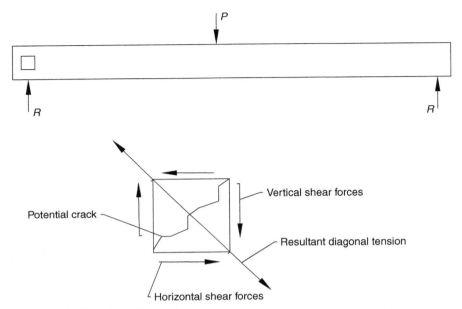

Figure 2.59 *Shear Near Bearing.*

direction of these stresses is perpendicular, or normal, to the face of this element. Shear stresses result from the deformation of the element by forces parallel to the face of the element (Figure 2.59).

Shear stresses develop in transversely loaded members as the externally applied forces tend to slide one section of a member past an adjacent section. Shear force tends to be at its greatest when adjacent to support locations of a beam, where the loads applied externally across the beam have to be countered by reaction forces equal in magnitude to the sum of the applied loads.

The general formula for shear stress in a beam is given by the first equation below. In practice, for beams with I or T shapes, this simplifies to the second equation shown. Unlike bending, the shear stress in a member for a given shear force depends on the area of the cross-section and not on the stiffness.

$$\tau = \frac{VQ}{It}$$

Where at a particular section, τ, maximum shear stress; V, shear force; Q, moment of area; I, moment of inertia; and t, thickness.

$$\tau_{average} = V / A_{web}$$

Where at a particular section, τ, average shear stress; V, shear force; and A, area of web.

A sudden extreme overload resulting in a shear failure may result in a clean vertical failure plane. However, the more common damage exhibited, particularly in a reinforced concrete beam, is a diagonal shear crack. Shear stresses on an element typically result in a plane of tensile stresses at 45° to the shear stresses. This results in cracks whose base points toward the bearing.

2.4.5 Fracture and Fatigue

Equations modeling the structural behavior for load effects such as tension and shear assume that stresses are smoothly distributed across the elements of the cross-section of a member. However, discontinuities in a member disrupt the flow of stresses and result in a rise in stress at that particular point. These discontinuities may include defects in the material, sharp changes in geometry, small cracks in the section, or the presence of large residual stresses due to operations such as welding. Excessive constraints on the cross-section of a member can also induce a rise in local stresses, such constraints can be imposed by multiple structural elements framing into the same location.

High local tensile stresses may result in the formation of surface cracks. The initial formation of a surface crack releases strain energy, which in a ductile material such as steel, is absorbed by energy dissipation due to plastic flow in the material adjacent to the crack tip. Cycles of high-tensile stresses will drive the growth of a crack until it propagates through the thickness of the material. This is the phenomena of fatigue. A through-thickness crack may still grow slowly, until it reaches a critical size. After this, it is liable to propagate suddenly through the entire section of an element as the release of strain energy is unable to be dissipated rapidly enough by plastic flow at the crack tip. This is known as fracture. Fracture mechanics is the study of crack propagation under such conditions.

High local compressive stresses result in crushing or buckling. Fracture and fatigue are driven by tensile stresses. Since concrete has little strength in tension, fracture and fatigue are then phenomena associated almost exclusively with steel for bridge construction.

For a crack in the edge of a plate under tension (Figure 2.60), it can be shown that the stress ahead of the crack and just outside of the plastic zone is described by the equation:

$$\sigma = K_I / \sqrt{2\pi r}$$

The parameter, K_I, is known as the stress-intensity factor. It is a function of the existing crack size, the applied stress, and the magnitude of the strain

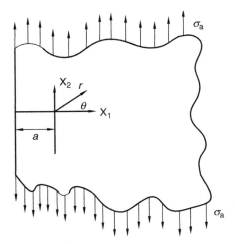

Figure 2.60 Edge Crack in Plate.

field. Different crack geometries have different functions for the stress–intensity factor. For the common case of an edge crack in a relatively large plate, the stress–intensity factor is:

$$K_I = 1.12\sigma_a \sqrt{\pi a}$$

For a crack of length $2a$ in the center of a relatively large plate, the stress-intensity factor is:

$$K_I = \sigma_a \sqrt{\pi a}$$

The rate of growth of a crack varies with the stress–intensity factor for each cycle of loading. Experiments show that the rate of crack growth per cycle of load may be divided into three distinct regions on a log–log plot (Figure 2.61).

Growth in regions I and III is sensitive to the microstructure of the material and is hard to predict. The parameter ΔK_{th} is a threshold value for the stress–intensity factor below which existing cracks will not grow. The parameter, K_{Ic}, is the critical stress–intensity factor. Values of the stress–intensity factor above K_{Ic} may propagate suddenly, resulting in fracture. The critical stress–intensity factor is a measure of the materials toughness, that is, its resistance to cracking. As stress–intensity factors are functions of applied tensile stress and crack size, for an expected range of stresses, a critical crack size can be calculated.

Crack growth in region II dominates most of the life of a crack. Crack growth can then be described by Paris' law:

$$\frac{da}{dN} = C(\Delta K)^m$$

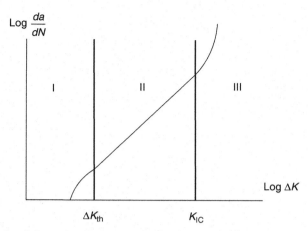

Figure 2.61 *Crack Growth per Cycle as a Function of Stress-Intensity Factor.*

Where, a, width of the crack; N, number of cycles of load; C and m, constants for the material, and ΔK, the range in the stress–intensity factor applied.

The constant, m, equals 3 for carbon steels. Since 1974, the AASHTO bridge design specification has organized specific connection and welding details on steel bridges into categories. Large-scale testing of these details has been used to develop stress range versus cycle to failure (S–N) curves. The number of loading cycles can be related to the traffic counts for a particular bridge providing a predicted fatigue for details (Figure 2.62).

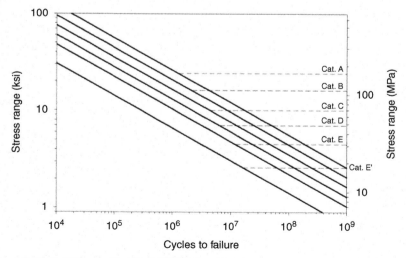

Figure 2.62 *S–N Curves Used in AASHTO Specifications.*

2.5 BRIDGE MATERIALS

2.5.1 Concrete and Reinforced Concrete

2.5.1.1 General Usage

The most commonly encountered material on almost any inventory of modern highway bridges is concrete. Well-made and well-placed concrete can provide good compressive strength as well as durability and can be formed into a wide variety of shapes. With the addition of steel bars or welded wire to provide tensile strength, almost any bridge element can be formed from reinforced concrete.

2.5.1.2 Properties

In general, the term concrete refers to a composite material composed of an aggregate (or filler) embedded in the hard matrix of a binding agent (a cement). The concrete used in the vast majority of applications in bridge construction is portland cement concrete, a concrete where the aggregates are rock and sand and the binder is portland cement paste. There are some specialty uses where the binder is an epoxy or elastomer or some similar polymer. Polymer concretes are more resistant to impact and are less permeable than portland cement concrete, but are also much more expensive. In this section, concrete will refer to portland cement concrete unless otherwise noted.

There are five basic constituents to concrete. Portland cement, water, and air combine to form the basic paste. Aggregates, varying in size from typically 1.5 in. in diameter down to sand, form the filler, constituting 70–80% of the volume of concrete. The fifth constituent is admixture. An admixture may or may not be used. Admixtures are either minerals or chemicals added to change the permeability, consistency, set time, or strength of the final concrete. Typical admixtures include silica fume to reduce permeability, plasticizers to increase the workability of concrete during placement, and retarders to increase the time before the concrete sets.

The concrete compressive strength used in bridge construction varies from 2500 psi for nonstructural items such as aprons for culverts to 6000 psi. Since the 1970s, 4000 psi has been the compressive strength commonly specified for structural use in bridge construction. Higher strengths are commonly used for prestressed concrete beams. Much higher strengths have been used for specialty applications. The shear and tensile strengths of concrete are much less than its compressive strength; shear is approximately 12% of the compressive strength and tensile strength is approximately 10%.

To compensate for the lack of tensile strength of concrete, almost all concrete bridge members use reinforcing steel to accommodate tensile stresses. Members such as culvert aprons may have a minimal amount of reinforcing, used to resist tensile stresses resulting from shrinkage of the concrete during curing or from settlement of the member. Columns, beams, girders, decks, and walls have significantly more reinforcing. Reinforcing steel used for structural members is usually in the form of round bars, 0.5–1.41 in. in diameter, with deformations on their surface to facilitate bond between the concrete and steel. Reinforcing steel used since the 1980s in highway bridge construction has typically had a yield strength of 60,000 psi. Prior to this, 40,000 psi was common. In older bridges, square-reinforcing bars may have been used.

Since the 1980s epoxy coated reinforcing steel has been widely – but not universally – used to protect against corrosion in applications where reinforcing steel near the surface of a member would likely be exposed to water. One such location is the deck of a bridge. Concrete is a porous material and is inherently permeable since the portland cement paste can never completely fill the matrix between aggregate particles.

Over the past couple of decades, reinforcing bars formed from fiber-reinforced polymers have come into use in bridge construction. These bars offer the advantage of not corroding when exposed to water or salt. However, this material is incapable of sustaining significant plastic deformation without fracture. This makes failure of members reinforced with these bars more sudden, and without the telltale of excessive deflection prior to complete loss of section.

Concrete deforms elastically over the range of stresses for which it is designed as do all materials used in bridge construction. However, sustained loading can result in additional deformations due to creep of the material. Deformations from creep are permanent, however, they do not directly cause damage to concrete. Creep may be most noticeable on long span reinforced concrete structures where excessive deflections may impact the rideability of the bridge.

2.5.1.3 Deterioration

Damage in concrete will generally manifest itself at the surface in one of three ways: cracking, scaling, or spalling.

A crack is simply a break in the concrete. The length of the crack is perpendicular to the tensile stresses that caused the crack. Cracking in concrete will appear either as pattern cracking (also referred to as map cracking) or as

Figure 2.63 *Map Cracking.*

individual cracks. Pattern cracking extends across the surface of a concrete member in two dimensions, appearing as a web of interconnected cracks. Alternately, it may be described as similar in appearance to the lines on a road map, hence the term map cracking (Figure 2.63). Pattern cracks occur due to restraint in the surface layer of the material relative to either shrinkage of the surface, or expansion of the volume of material underneath.

Individual cracks may be singular and isolated or appear in a pattern of a series of cracks. A series of cracks due to the same cause will, in general, be parallel in orientation and will occur at regular intervals. Whereas pattern cracks are limited to the surface of a concrete member, individual cracks may be superficial or may extend completely through a concrete member, depending on the cause of the cracking.

Cracks may result from structural issues or nonstructural issues. Structural cracks occur when a concrete member has been loaded past its elastic capacity (Figure 2.64). Such a loading may be one greater than anticipated in the original design, or the capacity of a member may have been reduced by wear, deterioration, or prior damage. Specific issues of structural cracking will be discussed with applicable remediation in Chapter 5.

Nonstructural cracking is usually due to processes that change the volume of the concrete and induce stresses within the member. This may be due to temperature, shrinkage, or chemical reactions within the concrete. Concrete expands and contracts in volume with rises and falls in temperature.

Figure 2.64 *Typical 45° Shear Crack in Concrete Girder.*

Temperature cracks can form when a concrete member is constructed and expansion due to heat of hydration during the curing process is not provided for, or when a member is thick enough that a significant temperature differential through the member during the curing process (Figure 2.65). Temperature cracking may also occur while the member is in service and the thermal movement of bridge members results in stresses in areas where

Figure 2.65 *Temperature Cracks.* *(Adapted from FHWA Bridge Inspectors Reference Manual [7]).*

the concrete is restrained due to poor details or when expansion joints fail to operate.

Shrinkage cracking occurs when concrete loses water due to evaporation (Figure 2.66). Shrinkage that occurs prior to the set of the concrete is plastic shrinkage cracking. These cracks typically appear within 24 h of placement and are parallel and spaced 1–3 ft. apart. This cracking is why it is important to maintain a wet cure in the hours and days after initial placement of concrete.

Even after the concrete has cured, it will continue to lose water and will therefore shrink. This loss of volume is unavoidable and can result in dry shrinkage cracks if a member is not properly provided with contraction joints. Plastic shrinkage cracking tends to be more of a concern to those involved with bridge maintenance because these cracks may extend a significant depth into a concrete member and provide a path for the intrusion of water and contaminants.

The concrete matrix is a highly alkaline environment capable of sustaining significant chemical reactions given the right conditions. Among those which are of most concern for their impact on the durability of concrete are: alkali–silica reactions, alkali–carbonate reactions, and delayed ettringite formation. The first two are reactions between alkalis in the cement paste and minerals in the aggregates of the concrete; the last is the development of a normally occurring reaction product in the concrete after the concrete

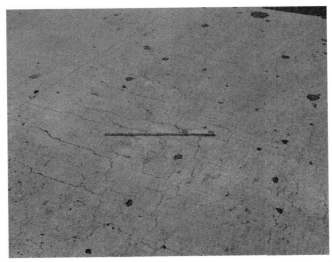

Figure 2.66 *Shrinkage Cracks. (Adapted from FHWA Bridge Inspectors Reference Manual [7]).*

Figure 2.67 *Alkali–Silica Reaction. (Adapted from FHWA Bridge Inspectors Reference Manual [7]).*

has already set due to heat and an excess availability of water for continued hydration. All three of these reactions result in the development of expansive products within the already hardened concrete matrix (Figure 2.67).

For reinforced concrete, corrosion of steel reinforcing is also a consideration. The normal protection against corrosion provided by the alkali environment of the concrete matrix can be disrupted by the presence of chlorides. Common sources of chloride exposure include deicing salts used on roadway and location in a marine environment. The products of steel corrosion have up to ten times the volume of the original steel. The corrosion and expansion of reinforcing steel is a common source of cracking and spalling of concrete in highway bridges.

Scaling is a breakdown of the concrete matrix over an area (Figure 2.68). Initially there is a loss of mortar, which as the scaling deepens, leads to the

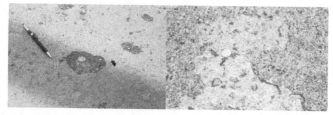

Figure 2.68 *Concrete Scaling, Light and Heavy. (Adapted from FHWA Bridge Inspectors Reference Manual [7]).*

loss of aggregate. Scaling can be driven by freezing and thawing, by the attack of acids (such as sulfuric acids resulting from combustion product from automobiles), and by wear from traffic.

Spalling is the loss of sections of the surface of a concrete member. Unlike scaling, where the surface is disintegrating, spalling is when a section falls off as an intact fragment. Spalls may result as heavily cracked surfaces lose their integrity or due to an overstress at the surface of the member. Most often this overstress is due to corrosion, and resulting expansion, of reinforcing steel in the member. It may also be due to restrained thermal movement, such as at a nonfunctioning expansion joint, or to an overload due to externally applied loads (Figure 2.69).

To facilitate bridge maintenance strategies, the causes for deterioration in concrete can be divided into three categories: load induced, environmental induced, and due to inherent deficiency.

Load-induced deterioration is the damage due to impacts from vehicular impacts, overstresses from loading that exceed the capacity of the member, and wear. Bridge management strategies to address these causes of deterioration include managing the routing of overheight and/or overweight trucks by requiring review and the assigning of permits for such vehicles. Wear can be mitigated by preventative maintenance actions such as the application of wearing surface overlays on bridge decks to provide a sacrificial cover over the element.

Most concrete deterioration resulting from environmental exposure is via water, by either the expansion of water in the freeze–thaw cycle, by contaminants carried in water into the member, or the combination of both effects. Preservation actions to maintain proper drainage of runoff and

Figure 2.69 *Concrete Spall. (Adapted from FHWA Bridge Inspectors Reference Manual [7]).*

to reduce the porosity of elements such as bridge decks are key to address this cause.

Deterioration from an inherent deficiency in either the design of the member or in the material properties of the concrete or its constituent components may be difficult to address. Bridge members designed with insufficient capacity for their expected loading (such as inadequate shear capacity in a concrete girder due to insufficient shear steel reinforcing) may require substantial maintenance work to add the needed capacity. Damage from material deficiencies may require substantial replacement of material.

2.5.2 Prestressed Concrete

2.5.2.1 General Usage

Since the 1950s in Europe and America, a form of concrete construction has been used in bridge building where steel reinforcing of concrete members is pulled into tension to induce compressive stress in the concrete. Applying a prestressing force at the bottom of a simple beam counteracts the natural state of tension, resulting in an increase in the load–carrying capacity of the beam (Figure 2.70).

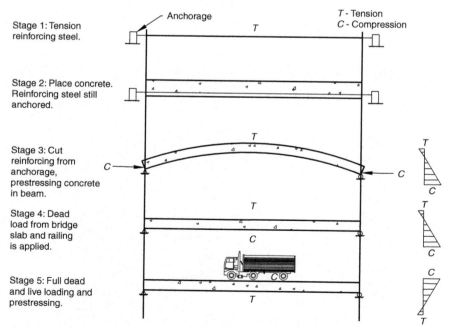

Figure 2.70 Effect of Prestressing.

Figure 2.71 *Construction of Prestressed Concrete Inverted-T Beam Bridge.*

Since prestressing is used to counter tension resulting from flexure, the bridge members most commonly prestressed are beams, girders, and the spanning members of other superstructure configurations. Deck panels which span between beams lines are another common prestressed concrete member. Prestressing is also used to form concrete piles, which can resist the bending stresses induced by pile driving.

The prestressing tension may be applied either before or after concrete member is formed and cast. Pretensioning is used for members that are precast by fabricators. Deck panels, piles, and beams are examples of pretensioned members. Typically beams are shaped like a capital I for efficiency, but shapes like an inverted capital T and planks are also available. The latter shapes have advantages in speed of construction (Figure 2.71).

Tension applied in the field is posttensioning (Figure 2.72). It may be done to a member that has been cast and set, such as a structural slab or a

Figure 2.72 *Posttension Ducts in Haunched Slab.*

cast-in-place box girder; or used to splice together precast member, such as a segmental box girder.

2.5.2.2 Properties

The concrete and the reinforcing steel used in prestressed concrete members is usually higher in strength than that commonly used for reinforced concrete. Common concrete compressive strengths used in prestressed members range from 5,000 psi to 8,000 psi, with considerably higher (more than 12,000 psi) used on occasion. Reinforcing consists of mild steel (plain carbon steel) bars used for shear reinforcing and to provide capacity against stresses from handling the beams. Prestressing is provided by high-strength steel, high-carbon steel wires, strands, or bars. Wires are typically 0.25-in. in diameter with a yield strength of 270,000 psi. Strands are bundles of wire with seven wire grade 270 strands being common. Bars are typically hot rolled, high-strength bars with an ultimate strength of 145,000 psi.

By definition, prestressed concrete is under a sustained load from the compression induced by the applied pre- or posttensioning. This will cause creep in the concrete and will also cause relaxation in the prestressing reinforcing. For minimum elongation under load, the prestressing reinforcing used is typically low-relaxation material for wires and strands and bars are formed by hot rolling as opposed to being cold drawn. Creep in the concrete is minimized by using higher-strength concretes. Higher-strength concrete typically has a higher modulus of elasticity and is stiffer and more resistant to creep.

2.5.2.3 Deterioration

Prestressed concrete is subject to the same mechanisms of deterioration as reinforced concrete. However, due to the presence of constant compressive stress, a prestressed concrete member should be less permeable, and therefore less vulnerable to damage from the environment than either plain or reinforced concrete members.

Prestressed members may be vulnerable to corrosion of the reinforcing steel near the anchorages for the steel. The very ends of a prestressed concrete beam are not in compression and typically have significant amounts of mild steel reinforcing.

2.5.3 Steel

2.5.3.1 General Usage

Along with concrete, steel is the other material found in almost all highway bridge construction. As discussed earlier, it is used in combination with plain concrete to form reinforced and prestressed concrete members.

It is also used extensively on its own to form bridge members. The stock of material used to form structural members is available in the form of plates, bars, and rolled shapes. The individual pieces of stock material are joined by welding, bolting, or (in construction prior to approximately 1960) riveting.

The most conspicuous steel members are those comprising the framing of the superstructure. The most common steel superstructures are girders, formed from either rolled W shapes or from plate girders. Long span trusses, arches, and complex bridge superstructures are almost exclusively constructed with steel members. Steel is also almost always used to form the secondary framing in the superstructure for bracing and load distribution.

Steel piling is commonly used as the foundation element to support bridge substructures (Figure 2.73). On some bridges, the piles extend above the ground line up to the pier cap to form the vertical supporting element of the pier.

In multiple, miscellaneous applications where higher-strength material is required: cables, tension rods, bearing devices, expansion joints, etc. steel has been the material used (Figure 2.74).

2.5.3.2 Properties

Prior to the use of Bessemer process, which allowed for the economical production of steel on a large scale, the metal used in bridge construction was iron. Cast iron, which contains 2–4% carbon, and wrought iron, which

Figure 2.73 *Steel Pile Bent.*

Figure 2.74 *Steel Cables for Network Arch.*

has less than 0.05% carbon and has inclusions of slag rolled into a fibrous structure, were used. Cast iron has a high compressive strength but is brittle. Wrought iron has good ductility and tensile strength, particularly along the grain of the fibers.

Steel is also an alloy of iron and carbon. Its carbon content is between that of cast iron and wrought iron. The processes to make steel allow for precise control of the carbon content and the content of small amounts of additional elements such as manganese. Steel has good tensile and compressive strengths in all directions and is ductile. Several variations in the precise composition of steel have been developed over the years and continue to be developed. Typically, the tensile and compressive strengths of steel have improved as has its resistance to corrosion and its toughness, that is, the ability to resist fracture. The steel commonly used in the first part of the twentieth century, designated as A7 by the American Society for Testing and Materials (ASTM), has a yield strength of 33,000 psi; while contemporary A709 steel is available in grades of 36, 50, 70, and 100.

Steel with a yield strength of 36,000 psi has been widely used since the 1960s with A36 steel. Today the use of grade 50 steel is the most common for use in fabricating the primary members for bridges since the cost is very similar to grade 36. Grade 36 is commonly specified for secondary members. The higher strengths are used on occasion for long span structures to minimize the steel required. The modulus of elasticity of steel is typically assumed as 29,000,000 psi for all grades and designations of steel. This means that the deflection under load of a member with a given cross-section is the same whether it is formed from A7 or from A709 grade 100. Since bridges must maintain limits on deflection to maintain serviceability, this limits the potential reduction in section (and therefore material) from specifying a high-grade material.

For bridge maintenance, there are three properties of interest that vary among types of steel: weldability, corrosion resistance, and fracture toughness.

The carbon content of steel plays a key role in determining whether it may be welded without having a propensity to cold crack. The higher the carbon content, the harder the steel is. The process of welding inevitability induces residual stresses in the steel near the weld, an area called the heat-affected zone. In a harder steel these residual stresses may induce cracking as the heat-affected zone cools or may reduce the available capacity of the steel to carry load such that cracking can occur during the normal service of the structure.

Contemporary structural steels have low carbon contents to address this. The carbon content of older steels, when members were assembled by riveting, may be too high to safely weld or to weld without requiring extensive pre- and posttreatment during the welding process. Elements other than carbon, such as manganese and silicon, used to formulate steel have a similar, but lesser effect on hardness. If an older steel bridge is a candidate for repair or retrofit by welding, a chemical test of the steel should be done to determine the composition of the existing steel alloy.

The American Welding Society gives the following formula in Annex I of the 2010 edition of their D1.1 Structural Welding Code for an equivalent carbon content:

$$CE = C + \frac{Mn + Si}{6} + \frac{Cr + Mo + V}{5} + \frac{Ni + Cu}{15}$$

Where all terms are in percentage of mass. CE = carbon equivalent and the terms on the right side of the equation are the various elements. Steels with CE < 0.30% may be welded without special considerations. Steels with CE > 0.50% are not readily weldable.

Adjusting the percentages of constituent elements can also affect the ability of the steel to resist corrosion. Stainless steel is formed by significantly increasing the amount of chromium to 8% or more. The amount of nickel may be increased as well, but it is the layer of chromium oxide that forms on the surface that provides protection against corrosion to the underlying material. The high levels of chromium negatively affect the weldability of stainless steel. The high cost of material and fabrication relative to typical structural steels preclude stainless steel from being an economical choice for forming bridge members; however, it is used in some specialty applications such as bearings.

Weathering steel is an alloy formulation that forms a tight patina of rust on its surface with exposure to the environment. The patina seals the interior of the steel member and protects it against any further corrosion. The contemporary designation used for weathering steel used in highway bridges is A709W. The previous alloy used starting in the mid-1960s was A588. Properly used, weathering steel can result in lower maintenance costs by eliminating the need for painting. However, if weathering steel is used in an environment that is constantly humid and has a source of chlorides (i.e., a marine environment, or a confined area where salt-laden runoff from an adjacent roadway is present) the weathering reaction may continue past formation of the patina and result in significant corrosion.

Fracture toughness is a material's ability to resist crack propagation. The critical stress–intensity factor of steel, K_{Ic}, can be determined through compact tension tests per ASTM specification E399. However, these tests can be expensive to conduct. The standard way to specify fracture toughness for steel is in terms of Charpy impact test results.

In a Charpy impact test, a standardized hammer on a pendulum is raised then released to break through a standard notched coupon of material. The height reached by the hammer after it travels through the specimen indicates the energy absorbed by the material in the break. The more energy absorbed, the tougher the material is. There is a reliable correlation between Charpy results and the critical stress–intensity factor and the Charpy test is simple and cheap enough to provide effective quality control. If there are concerns about the fracture toughness of an existing bridge, a coupon of steel can be taken for testing. The standard specimen size for a Charpy test is only $10 \times 10 \times 55$ mm and test provisions for smaller sizes are available.

The fracture toughness of steel is affected by its temperature. The lower the temperature, the lower the toughness. This, combined with stresses induced by the thermal contraction of a bridge in cold weather, can lead to cracking at locations that are overly restrained, even without prior visible fatigue cracking. This is what occurred on the Hoan Bridge on the morning of December 13, 2000, in Milwaukee, Wisconsin. Three 10-ft. deep girders fractured, two completely through, where the webs were highly constrained by the connection to the lateral bracing (Figure 2.75).

Figure 2.75 *Triaxial Constraint Condition on Hoan Bridge. (FHWA Inspection Manual [7]).*

2.5.3.3 Deterioration

As a material, the mechanism responsible for the deterioration of steel is corrosion. As a chemical process, the reaction can be described by the equation:

$$4\mathrm{Fe} + 3\mathrm{O}_2 + 2\mathrm{H}_2\mathrm{O} = 2\mathrm{Fe}_2\mathrm{O}_3\mathrm{H}_2\mathrm{O}$$

Oxygen from the atmosphere combines with water and with iron in the steel to form hydrated ferric oxide, otherwise known as rust. The area of rust has negligible strength and does not carry load, reducing the effective cross-section of the member where corroded.

Rust initially appears as a discoloration, but the corrosion products are several times the volume of steel. As the corrosion expands, the surface will begin to pit, then form flakes, which stratify into layers of scale as the corrosion penetrates into the member (Figure 2.76). It may result in complete loss of section if allowed to continue.

Corrosion of steel occurs only in the presence of both oxygen and water and is hence typically driven by wet and dry cycles of exposure. Corrosion is accelerated in the presence of chlorides, typically available on highway bridges from deicing salts used to clear roadways or some exposure to a marine environment.

Other than visible loss of section, the two primary indicators of the deterioration of steel in a structure are cracking and buckling. Both may be

Figure 2.76 *Corrosion Through Web of Steel Beam.*

caused by the overloading of a steel member. However, the gross shear capacity of a steel member is very rarely exceeded before reaching failure in bending, which would result in plastic deformation of the section and buckling of the compressive side of the beam. Cracks in steel in bridges are primarily due to fracture or fatigue. These will be found to initiate at areas of restraint or discontinuities in the member's cross-section. Fatigue cracks will begin small and tight and may grow with cycles of loading (Figure 2.77).

A compression member or portion of a bending member subject to compressive stresses will twist and sweep out of plane when its buckling capacity is exceeded. This may happen due to a gross overload on the structure or an impact to the structure, or it may happen when the capacity at that point on the member is compromised by loss of section or by exposure to fire. The high temperatures from fire can lower the yield strength of steel.

2.5.4 Timber
2.5.4.1 General Usage
Concrete and steel construction dominates modern highway bridge construction and comprises the vast majority of the existing inventory of bridges on public roads. Prior to the wide availability of steel, the material used for flexural strength was wood. Truss and beam superstructures were

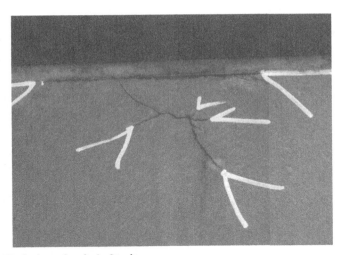

Figure 2.77 *Fatigue Cracks in Steel.*

constructed with wood. Note, the term timber is used to refer to wood prior to the point in processing where it is cut into boards as nominal lumber. Nominal lumber pieces are planed down to precise widths and thicknesses to form dimensional lumber. The pieces of wood which are used for primary members in bridge construction are larger than boards and are referred to as timbers (rectangular shapes) or posts (round shape); hence, the convention is to refer to wood bridge construction as timber construction.

According to the 2013 NBI, 3.5% of all bridges on public roads in United States are timber. This number, however, does not count bridges with steel beam superstructures on timber abutments or pile bents. Nor does it include structures with a length of less than 20 ft. Bridge maintenance activities on a large number of inventories will involve dealing with bridges or smaller structures, which use timber as a structural component. Often these bridges may be some of the older ones in the inventory. The last half of the 1930s were the peak years of construction for those timber bridges that were on the 2013 NBI. Timber bridges will usually be found in locations which are less demanding on bridges, such as lower traffic volume roads, as bridges for right of way access, and in parks or other remote areas.

Timber bents are an element common to timber bridges, and to bridges using steel beams on timber substructures. The bents may be used as piers or, with lagging to retain fil, as timber abutments. Timber bents with steel beam spans allowed for relatively quick construction of small span bridges without requiring heavy equipment (Figure 2.78).

Figure 2.78 *Timber Abutment Bent.*

Spans of 25 ft. or less may be framed with timber beams (Figure 2.79). Decks may be formed with lumber planks.

Timber formed from a single piece of wood is referred to as solid sawn. Newer bridge members may be fabricated by a process of using layers of wood joined by adhesive to allow for consistent and improved performance of the material. These members are referred to as glued laminate (also referred to as glulam) members.

Figure 2.79 *Short Span Timber Bridge.*

Table 2.1 Comparative allowable stresses for timber construction

Loading	Structural grade Douglas fir	Structural glued laminated softwood (20F-V1 DF/WW)	KDOT recommendation for unknown species
Bending, parallel to the grain	1500	2000	1200
Compression, parallel to the grain	1100	1000	850
Compression, perpendicular to the grain	625	650	400

Source: From KDOT, Bridge Design Manual [2]

2.5.4.2 Properties

Whether solid sawn or glulam, the capacity of wood members varies significantly according to the orientation of loading. Wood is a fibrous material. Reference is made to whether loading is parallel to the fibers (along the grain) or perpendicular to the fibers (against the grain). In solid sawn wood, strength also varies by species of wood, by grade (quality) of wood, and even slightly between dimensions of wood pieces. Table 2.1 shows the allowable stresses for a timber formed from new structural grade Douglas Fir, glulam, and recommended allowable stresses when the species and grade on in-place timber is unknown.

2.5.4.3 Deterioration

In wood, the analog to cracks are checks, splits, and shakes (Figure 2.80). Checks are separations of the wood fibers parallel to the grain. Splitting is

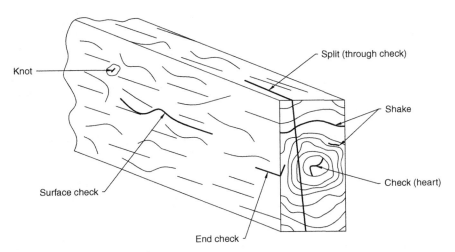

Figure 2.80 *Defects in Timber.*

a separation parallel to the grain that extends completely through the piece. Shakes are separations of the fiber occurring between the annual growth rings.

These defects may be present in the wood member from the time it was initially kiln–dried, caused by differential shrinkage across the cross-section, or they may be incurred during the life of the bridge due to weathering or overloading of the member. Timber exposed to the weather loses moisture due to the drying actions of the sun and wind. The effects can be uneven across the member depending on its exposure. Existing checks in the surface also allow for water to penetrate and then expand upon freezing, further driving check growth. A member may be loaded beyond its capacity to respond without damage due to either an extreme load, or due to a lesser loading on a section with diminished capacity due to previous damage.

The primary cause of significant damage to timber bridge members is biological attack. This can take the form of decay caused by microorganisms (primarily fungi) or consumption by either insects or crustaceans. Many species of fungi can live on wood, but significant damage is caused by one of the two types: brown rot and white rot. Both attack the cell walls, resulting in a loss of integrity. Areas damaged by brown rot are brittle while areas damaged by white rot are soft. Multiple species of insects and crustaceans (including termites, carpenter ants, marine borers, and several others) bore extensively through wood, hollowing the interior of the member. Damage from biological attack may appear as a loss of section at the surface, or may appear as surface indentations from voids underneath.

2.5.5 Other Materials

2.5.5.1 Stone Masonry

There are fewer stone masonry bridges on the 2013 NBI than timber bridges, with less than 1700 reported making up less than 0.3% of inventory. The majority of these were built before 1940 with less than ten new masonry bridges constructed annually in every year after 1950. As with timber bridges, the NBI does not count structures with an opening of less than 20 ft., nor does it count the number of span bridges with a stone substructure element (Figure 2.81).

Where stone has been used, it has almost always been locally sourced. The mechanical and chemical properties vary with the type available from the local geology. The compressive strength of the rock used is almost always equal to or greater than concrete. The mortar used will typically be a type of portland cement and will be weaker than the rock.

Figure 2.81 *Stone Masonry Substructure.*

Stone masonry deteriorates by weathering (breaking into small particles), spalling (small sections of rock popping out), or splitting (separations occurring either at seams in the rock or along mortar joints). Possible causes include: chemical reactions, thermal expansion and contraction, freeze-thaw, plant growth in the seams, and movement of the soil mass behind stone masonry walls.

2.5.5.2 Fiber-Reinforced Polymer

Fiber-Reinforced Polymer (FRP) is a composite material formed from combining a matrix of polymer resin with either glass (GFRP) or carbon fibers for tensile strength. The polymer resin is usually polyester, though either a vinyl ester or epoxy may be used. FRP is light, has a high-tensile strength and can be manufactured in a number of forms. The common forms are fabric, strips, and rods. It may also be pultruded in structural shapes.

FRP fabric and strips are used for repairs and retrofits of damaged or deficient members to provide tensile capacity. The technology was an extension of the prior technique of externally bonding steel plates to deficient concrete members. FRP retrofits came into use in the United States in the 1990s. The first AASHTO Guide Specification for the Design of Bonded FRP Systems for Repair and Strengthening of Concrete Bridge Elements was published in 2012 (Figure 2.82).

For new construction, GFRP reinforcing bars are available for use as an alternative to steel. The appeal to the use of FRP as a replacement for steel is

Figure 2.82 *Repair of Prestressed Concrete Beam With FRP.*

that FRP does not corrode with exposure to water and chlorides. This use, too, has a new AASHTO specification with the 2009 AASHTO Load and Resistance Factor Design (LRFD) Bridge Design Guide Specifications for GRRP-Reinforced Concrete Bridge Decks and Traffic Railings.

Because of its corrosion resistance to chlorides, several demonstration projects have been implemented using FRP deck panels. The light weight of these deck panels also offers the promise of improving load ratings on bridges through a reduction of dead load. Though a number of FRP decks are in service, the technology has not been generally adopted for use. There exist issues of initial cost, connection detailing between FRP and steel and concrete members, and familiarity with the material.

REFERENCES

[1] McGrath TJ, Beaver JL, Timothy J. Management of Utah Highway Culverts. J Transport Res Board 2005;1904:113–23.

[2] Kansas Department of Transportation. Bridge design manual. Topeka, KS: State of Kansas; 2014.

[3] Briaud JL, James RW, Hoffman SB. Settlement of bridge approaches (The bump at the end of the bridge). NCHRP Synthesis of Highway Practice 234. Washington, DC: Transportation Research Board; 1997.

[4] Federal Highway Administration. Bridge maintenance training. Washington, DC: US Department of Transportation; 2003.

[5] Federal Highway Administration. Geosynthetic reinforced soil integrated bridge system synthesis report. Washington, DC: US Department of Transportation; 2011. FHWA-HRT-11-027.

[6] Kansas Department of Transportation. Bridge construction manual. Topeka, KS: State of Kansas; 2012.

[7] Federal Highway Administration. Bridge Inspector's Reference Manual. Washington, DC: US Department of Transportation; 2012. FHWA NHI 12-049.

CHAPTER 3

Bridge Inspection and Evaluation

Overview

The cornerstone of bridge preservation is inspection. This chapter discusses the current bridge inspection process in the United States as relevant to scoping and programming bridge maintenance work. Comparisons are made to bridge inspections in other countries. Specific inspection methods that provide information for scoping bridge repair and rehabilitation work are considered. Bridge load rating and evaluation and their relation to bridge maintenance are discussed.

3.1 INTRODUCTION

The start of the bridge preservation process is inspection. The inspector's observations provide the basis for subsequent assessment and evaluation of a bridge. The inspector is the first line of safety, acting to close structures when dangerously deficient conditions are found.

This text is not intended to be a comprehensive guide to conducting bridge inspections. The *FHWA Bridge Inspector's Reference Manual* discusses the requirements for bridge inspections in the United States. Individual state departments of transportation also have published guidelines for the conduct of bridge inspections within their respective states. Bridge inspections are discussed in this text with an eye to their use in scoping bridge maintenance activities. Bridge inspections provide data for:

- An accurate description of a bridge as built and as it currently exists. Plans that may be on file from the original contract letting for construction of a bridge only reflect the intent of the bridge designer. It is not common for the bridge "as-built" to have been modified slightly in geometry due to construction errors or to conditions encountered in the field. Also, changes made during maintenance activities subsequent to initial construction, such as the replacement of expansion joints, are not always reflected in plans on file for a bridge.
- Alerting a bridge owner to damage that has occurred to a structure. Bridges are subject to damage from incidents by traffic (e.g., impacts for vehicles) or from the environment (e.g., scour after heavy rainfall events).
- Assessment of the declining condition of bridge members due to corrosion or other long-term deterioration of materials.

Highway Bridge Maintenance Planning and Scheduling
http://dx.doi.org/10.1016/B978-0-12-802069-2.00003-9

Should the condition of a bridge member be found to have significantly deteriorated since the bridge's last inspection, the structural capacity of the bridge is evaluated to determine if it still has the capacity to carry the loads it is required to. For a highway bridge, those are traffic loads (the largest of which are from trucks carrying freight) as permitted on public roads by the laws of the jurisdiction in which the bridge is located.

These evaluations are referred to as load ratings. A load rating that requires closing a bridge or restricting the use of a bridge from heavier truck traffic is often the impetus for a maintenance action to restore is capacity to some degree. Load rating and its relation to scheduling maintenance activities will be discussed later in this chapter.

3.2 BRIDGE INSPECTION IN THE UNITED STATES

3.2.1 Types of Inspection

The National Bridge Inspection Standards (NBIS) for the United States are enumerated in law in Section 650 of Title 23 of the United States Code. All bridges on public roads in the United States are required to be inspected as per the standards and frequency, and by personnel qualified as outlined, in the NBIS. The NBIS established that the Department of Transportation in each state is responsible for the inspection and evaluation of all bridges on public roads within its borders, except those owned and maintained by Federal agencies, such as, the US Park Service.

Five types of inspection, each with their own scope, are described in the NBIS. They are:
* In-depth;
* Routine;
* Fracture critical (FC);
* Underwater; and
* Damage.

Except for damage inspections, each inspection must be led in the field by a team leader who has passed the FHWA Bridge Inspection course and who possesses a minimum number of years of experience, the requirements of which vary depending on education and possession of engineering licensure (Table 3.1).

In-depth inspections are close-up visual inspections of the bridge members. Although the frequency of this level of inspection is not prescribed in the NBIS, as a practical matter this type of inspection is required for the initial inspection when the Structure Inventory and Appraisal (SI&A) sheet is

Table 3.1 Requirements to be a team leader for bridge inspections

Min. number of years experience	Education	Licensure or certification	
None	Bachelor's degree★	Licensed professional engineer (PE)	All team leaders must have completed a FHWA approved comprehensive bridge inspection training course.
2	Bachelor's degree	Intern engineer (IE)	
4	Associate's degree in engineering or engineering technology	None	
5	None	Level III or IV Bridge Safety Inspector Certification under the National Certification in Engineering Technologies	
10	None	None	

★ A bachelor's degree is not explicitly required in the Federal Regulations for a licensed PE, but the requirements for a PE in the United States include a 4-year degree.

completed for each bridge (Figure 3.1). The SI&A is a physical description of the bridge and its condition. It forms the basis of the bridge inventory files required to be maintained by each state and of the NBI maintained by the FHWA. Any physical change to a bridge, occurs during a reconstruction or with substantial repair work, necessitate an update to the SI&A for the bridge and typically trigger a new in-depth inspection. States are also free to set additional criteria to initiate in-depth inspections.

Routine inspections are those regularly scheduled inspections to ascertain changes in the condition of bridges. They occur at intervals that are, unless specifically excepted, not to exceed 2 years. Bridges that are in danger of experiencing a quicker than usual drop in condition of a critical member or members may be inspected at a shorter frequency. An example of this would be a reinforced concrete girder bridge with shear cracking in the girders set at a 1-year inspection interval, with the intent of monitoring any crack growth. These inspections are not required to be close-up, but may be conducted at a distance that allows the inspector to detect any change in condition of members being examined. So an inspector may observe the

OMB No. 2125-0501

Structure Inventory and Appraisal Sheet

NATIONAL BRIDGE INVENTORY · · · · · · STRUCTURE INVENTORY AND APPRAISAL   10/15/94

```
********** IDENTIFICATION **********************
(1) STATE NAME      _____     CODE __
(8) STRUCTURE NUMBER          #_____
(5) INVENTORY ROUTE (ON/UNDER) - ___ = _____
(2) HIGHWAY AGENCY DISTRICT
(3) COUNTY CODE    ___   (4) PLACE CODE    ___
(6) FEATURES INTERSECTED - _____
(7) FACILITY CARRIED    - _____
(9) LOCATION            - _____
(11) MILEPOINT/KILOMETERPOINT        _____.__
(12) BASE HIGHWAY NETWORK -        CODE __
(13) LRS INVENTORY ROUTE & SUBROUTE   #_____ __
(16) LATITUDE          __ DEG __ MIN __.__ SEC
(17) LONGITUDE         __ DEG __ MIN __.__ SEC
(98) BORDER BRIDGE STATE CODE   __ % SHARE __ %
(99) BORDER BRIDGE STRUCTURE NO.   #_____

********** STRUCTURE TYPE AND MATERIAL **********
(43) STRUCTURE TYPE MAIN:  MATERIAL - _____
         TYPE - _____   CODE __
(44) STRUCTURE TYPE APPR:  MATERIAL - _____
         TYPE - _____   CODE __
(45) NUMBER OF SPANS IN MAIN UNIT         ___
(46) NUMBER OF APPROACH SPANS             ___
(107) DECK STRUCTURE TYPE -          CODE _
(108) WEARING SURFACE / PROTECTIVE SYSTEM:
   A) TYPE OF WEARING SURFACE  - _____  CODE _
   B) TYPE OF MEMBRANE         - _____  CODE _
   C) TYPE OF DECK PROTECTION - _____  CODE _

********** AGE AND SERVICE ********************
(27) YEAR BUILT                           ___
(106) YEAR RECONSTRUCTED                   ___
(42) TYPE OF SERVICE: ON - _____
         UNDER - _____   CODE __
(28) LANES: ON STRUCTURE __  UNDER STRUCTURE __
(29) AVERAGE DAILY TRAFFIC             _____
(30) YEAR OF ADT ____  (109) TRUCK ADT ___ %
(19) BYPASS, DETOUR LENGTH             ___ KM

********** GEOMETRIC DATA ********************
(48) LENGTH OF MAXIMUM SPAN              ___.__ M
(49) STRUCTURE LENGTH                    ___.__ M
(50) CURB OR SIDEWALK:  LEFT __.__ M  RIGHT __.__ M
(51) BRIDGE ROADWAY WIDTH CURB TO CURB   ___.__ M
(52) DECK WIDTH OUT TO OUT               ___.__ M
(32) APPROACH ROADWAY WIDTH (W/SHOULDERS) ___.__ M
(33) BRIDGE MEDIAN -           CODE __
(34) SKEW    __ DEG   (35) STRUCTURE FLARED __
(10) INVENTORY ROUTE MIN VERT CLEAR      __.__ M
(47) INVENTORY ROUTE TOTAL HORIZ CLEAR   __.__ M
(53) MIN VERT CLEAR OVER BRIDGE RDWY     __.__ M
(54) MIN VERT UNDERCLEAR     REF - ___  __.__ M
(55) MIN LAT UNDERCLEAR RT   REF - ___  __.__ M
(56) MIN LAT UNDERCLEAR LT              __.__ M

********** NAVIGATION DATA ********************
(38) NAVIGATION CONTROL - _____   CODE _
(111) PIER PROTECTION - _____   CODE _
(39) NAVIGATION VERTICAL CLEARANCE      __.__ M
(116) VERT-LIFT BRIDGE NAV MIN VERT CLEAR __.__ M
(40) NAVIGATION HORIZONTAL CLEARANCE    __.__ M
```

```
SUFFICIENCY RATING = ____._
STATUS = _____

********** CLASSIFICATION **************** CODE
(112) NBIS BRIDGE LENGTH - _____  __
(104) HIGHWAY SYSTEM      - _____  __
(26) FUNCTIONAL CLASS     - _____  __
(100) DEFENSE HIGHWAY      - _____  __
(101) PARALLEL STRUCTURE   - _____  __
(102) DIRECTION OF TRAFFIC - _____  __
(103) TEMPORARY STRUCTURE  - _____  __
(105) FEDERAL LANDS HIGHWAYS - _____  __
(110) DESIGNATED NATIONAL NETWORK - _____  __
(20) TOLL                 - _____  __
(21) MAINTAIN -             _____  __
(22) OWNER    -             _____  __
(37) HISTORICAL SIGNIFICANCE - _____  __

********** CONDITION ****************** CODE
(58) DECK                                    _
(59) SUPERSTRUCTURE                          _
(60) SUBSTRUCTURE                            _
(61) CHANNEL & CHANNEL PROTECTION            _
(62) CULVERTS                                _

********** LOAD RATING AND POSTING ******** CODE
(31) DESIGN LOAD      -         OR  _____  _
(63) OPERATING RATING METHOD - _____  _
(64) OPERATING RATING - _____  __._
(65) INVENTORY RATING METHOD - _____  _
(66) INVENTORY RATING - _____  __._
(70) BRIDGE POSTING  -                    _
(41) STRUCTURE OPEN, POSTED OR CLOSED - __  _
         DESCRIPTION - _____

********** APPRAISAL ****************** CODE
(67) STRUCTURAL EVALUATION                   _
(68) DECK GEOMETRY                           _
(69) UNDERCLEARANCES, VERTICAL & HORIZONTAL  _
(71) WATERWAY ADEQUACY                       _
(72) APPROACH ROADWAY ALIGNMENT              _
(36) TRAFFIC SAFETY FEATURES                 _
(113) SCOUR CRITICAL BRIDGES                 _

********** PROPOSED IMPROVEMENTS *************** 
(75) TYPE OF WORK -                      CODE __
(76) LENGTH OF STRUCTURE IMPROVEMENT     ___.__ M
(94) BRIDGE IMPROVEMENT COST     $___,___,000
(95) ROADWAY IMPROVEMENT COST    $___,___,000
(96) TOTAL PROJECT COST          $___,___,000
(97) YEAR OF IMPROVEMENT COST ESTIMATE   ____
(114) FUTURE ADT                       _____
(115) YEAR OF FUTURE ADT                 ____

********** INSPECTIONS *******************
(90) INSPECTION DATE __/__  (91) FREQUENCY __ MO
(92) CRITICAL FEATURE INSPECTION:     (93) CFI DATE
   A) FRACTURE CRIT DETAIL - __ - __ MO  A) __/__
   B) UNDERWATER INSP     - __ - __ MO  B) __/__
   C) OTHER SPECIAL INSP  - __ - __ MO  C) __/__
```

Figure 3.1 *Example SI&A Sheet.*

underside of a bridge at a tall grade separation from the ground. A bridge owner may petition the FHWA to allow for a longer bridge inspection interval of up to 4 years. This may be granted if it can be shown that there is a low risk associated with allowing such an interval. A buried reinforced concrete culvert in good condition on a low volume road might qualify for such an exception.

FC inspections are inspections of steel members loaded in tension whose loss would cause collapse of a bridge, in whole or in part. These inspections must be "hands on," conducted at no more than an arm's length from the observed FC member. They are conducted at intervals of 2 years or less. As a practical matter the inspection of steel members for cracks that may not be visible to the naked eye at their inception requires the use of nondestruction evaluation (NDE) techniques such as ultrasound or magnetic particle testing in FC inspections. NDE techniques are discussed later in this chapter.

Underwater inspections are inspections of the components of the bridge substructure that are not visible during low water elevations and of the topography of the channel adjacent to the substructure. These inspections focus on changes to bridge condition due to scour. Scour is the loss of material in the streambed due to the flow of water and can result in exposing substructure elements. If scour is severe it may result in a loss of capacity in foundation members such as piling and spread footings. Scour is one of the leading causes of bridge failures in the United States. Underwater inspections are required at a maximum interval of 5 years. Bridges with foundations that are particularly vulnerable to the effects of scour may be deemed scour critical and be placed on a shorter interval of inspection. Since increasing scouring may be triggered by large rainfall events, plans of action are required for each scour critical bridge. These plans of action include requirements to observe bridges after rainfall events of a particular intensity in the bridge's watershed and pre-existing plans to route traffic away from a bridge in the event that its stability cannot be confirmed.

Damage inspections are those unscheduled inspections to review the condition of a structure following a potentially damaging event. This could be the review of a bridge after an environmental incident such as an earthquake or a hurricane, or after a man-made incident such as a traffic impact or a fire.

One of the documents that the FHWA refers to as a reference for conducting bridge inspections, The *Manual for Bridge Evaluation* by American Association of State Highway and Transportation Officials (AASHTO), specifically lists three other types of inspection in addition to the five discussed earlier. These are initial, routine wading, and special. As discussed earlier, an initial inspection is typically an in-depth inspection conducted to fill out the SI&A sheet. Similarly, AASHTO recommends that a routine wading investigation be done concurrently with the routine inspection. Such an investigation involves probing the channel bed adjacent to substructure to check for any scour effects. As a practical matter such work involves entering and exiting the water and requires appropriate weather. Often routine wading inspections are scheduled on a 2-year cycle, but separate from routine inspections. Special inspections are those scheduled by an agency to monitor a particular situation on a particular bridge. An example would be annual monitoring of the condition of a bridge with a load restriction. The qualification of the inspector for a particular special inspection is determined by the state and should be commensurate with the investigation required.

The types of inspections outlined in the NBIS and the *Manual for Bridge Evaluation* are summarized in Table 3.2.

Table 3.2 Types of bridge insp ections

Type	Maximum interval	FHWA-qualified team leader required?	Close-up or hands-on inspection required?	Focus of inspection
In–depth	Needed with any significant change to the bridge.	Yes	Yes	Determining the condition of bridge members when a greater thoroughness is required than that used for routine inspections.
Routine	2 years	Yes	No	Determining any changes in condition from the previous inspection.
FC	2 years	Yes	Yes	Members specified as FC.
Underwater	5 years	Yes	No	Substructure below the water line and the adjacent streambed.
Damage	–	No	No	Damage from environmental or manmade incidents.
Initial	–	Yes	Yes	A complete assessment of the initial condition of the bridge for the SI&A sheet.

Table 3.2 Types of bridge inspections *(cont.)*

Type	Maximum interval	FHWA-qualified team leader required?	Close-up or hands-on inspection required?	Focus of inspection
Routine wading	2 years	Yes	No	Substructure and streambed below the water line that is accessible without diving equipment.
Special	–	Set by State for particular focus of inspection.	Set by State for particular focus of inspection.	Monitoring a particular area of concern that may affect the use and serviceability of the bridge.

3.2.2 Component Condition Ratings

Since the inception of the NBIS in 1971, after the collapse of the Silver Point Bridge, the state of each bridge in the NBI has been reported in terms of condition and appraisal ratings. Condition ratings describe the state of the components of a bridge relative to their condition as originally constructed (Table 3.3). For the SI&A, the components of a span bridge are defined as its deck, superstructure, and substructure while a culvert has a single condition rating reflecting the state of the entire structure. For bridges over waterways, a condition rating is assigned to the channel reflective of the stability of the channel of the waterway and any evidence of potential for erosion of the bank or scouring adjacent to the substructure of the bridge.

To assign a condition rating the state of a component is evaluated on a scale of 0–9. The structural condition ratings are meant to be an evaluation of the overall capacity of the entire component relative to its as-constructed capacity. The condition rating guidelines provided by FHWA are shown in Table 3.4.

Table 3.3 Condition ratings applicable by bridge type

Applicable condition rating	Bridge type
For span bridges	Deck, Superstructure, Substructure
For culverts	Culvert
For bridges over waterways	Channel

Table 3.4 FHWA component condition-rating guidelines

Rating	Deck/Superstructure/ Substructure	Culvert	Channel
N	Not Applicable	Not applicable. Use if structure is not a culvert.	Not applicable. Use when bridge is not over a waterway (channel).
9	Excellent Condition	No deficiencies	There are no noticeable or noteworthy deficiencies that affect the condition of the channel.
8	Very Good Condition – no problems noted	No noticeable or noteworthy deficiencies that affect the condition of the culvert. Insignificant scrape marks caused by drift.	Banks are protected or well vegetated. River control devices such as spur dikes and embankment protection are not required or are in a stable condition.
7	Good Condition – some minor problems	Shrinkage cracks, light scaling, and insignificant spalling that does not expose reinforcing steel. Insignificant damage caused by drift with no misalignment and not requiring corrective action. Some minor scouring has occurred near curtain walls, wingwalls, or pipes. Metal culverts have a smooth symmetrical curvature with superficial corrosion and no pitting.	Bank protection is in need of minor repairs. River control devices and embankment protection have a little minor deficiency. Banks and/or channel have minor amounts of drift.
6	Satisfactory Condition – structural elements show some minor deterioration.	Deterioration or initial disintegration, minor chloride contamination, cracking with some leaching, or spalls on concrete or masonry walls and slabs. Local minor scouring at curtain walls, wingwalls, or pipes. Metal culverts have a smooth curvature, nonsymmetrical shape, significant corrosion, or moderate pitting.	Bank is beginning to slump. River control devices and embankment protection have widespread minor deficiency. There is minor streambed movement evident. Debris is restricting the channel slightly.

5	Fair Condition – all primary structural elements are sound but may have minor section loss, cracking, spalling, or scour.	Moderate to major deterioration or disintegration, extensive cracking and leaching, or spalls on concrete or masonry walls and slabs. Minor settlement or misalignment. Noticeable scouring or erosion at curtain walls, wingwalls, or pipes. Metal culverts have significant distortion and deflection in one section, significant corrosion or deep pitting.	Bank protection is being eroded. River control devices and/or embankments have major deficiency. Trees and brush restrict the channel.
4	Poor Condition – advanced section loss, deterioration, spalling, or scour.	Large spalls, heavy scaling, wide cracks, considerable efflorescence, or opened construction joint permitting loss of backfill. Considerable settlement or misalignment. Considerable scouring or erosion at curtain walls, wingwalls, or pipes. Metal culverts have significant distortion and deflection throughout, extensive corrosion or deep pitting.	Bank and embankment protection is severely undermined. River control devices have severe deficiency. Large deposits of debris are in the channel.
3	Serious Condition – loss of section, deterioration, spalling, or scour have seriously affected primary structural components. Local failures are possible. Fatigue cracks in steel or shear cracks in concrete may be present.	Any condition described in Code 4 but which is excessive in scope. Severe movement or differential settlement of the segments, or loss of fill. Holes may exist in walls or slabs. Integral wingwalls nearly severed from culvert. Severe scour or erosion at curtain walls, wingwalls, or pipes. Metal culverts have extreme distortion and deflection in one section, extensive corrosion, or deep pitting with scattered perforations.	Bank protection has failed. River control devices have been destroyed. Streambed aggradation, degradation, or lateral movement has changed the channel to now threaten the bridge and/or approach roadway.

(Continued)

Table 3.4 FHWA component condition-rating guidelines (cont.)

Rating	Deck/Superstructure/ Substructure	Culvert	Channel
2	Critical Condition – advanced deterioration of primary structural elements. Fatigue cracks in steel or shear cracks in concrete may be present or scour may have removed substructure support. Unless closely monitored it may be necessary to close the bridge until corrective action is taken.	Integral wingwalls collapsed, severe settlement of roadway due to loss of fill. Section of culvert may have failed and can no longer support embankment. Complete undermining at curtain walls and pipes. Corrective action required to maintain traffic. Metal culverts have extreme distortion and deflection throughout with extensive perforations due to corrosion.	The channel has changed to the extent that the bridge is near a state of collapse.
1	"Imminent" Failure Condition – major deterioration or section loss present in critical structural components, or obvious vertical or horizontal movement affecting structure stability. Bridge is closed to traffic but corrective action may put bridge back in light service.	Bridge closed. Corrective action may put bridge back in light service.	Bridge closed because of channel failure. Corrective action may put bridge back in light service.
0	Failed Condition – out of service; beyond corrective action.	Bridge closed. Replacement necessary.	Bridge closed because of channel failure. Replacement necessary.

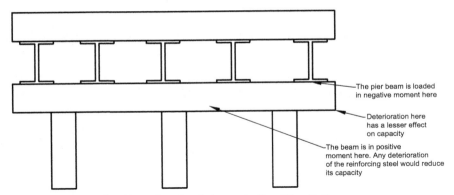

The pier beam is loaded in negative moment here

Deterioration here has a lesser effect on capacity

The beam is in positive moment here. Any deterioration of the reinforcing steel would reduce its capacity

Figure 3.2 *Example of Deterioration Relevant to Condition Rating.*

Localized deterioration that does not significantly impact a component's overall capacity should not be used to diminish its condition rating, although the deterioration should be reported in the routine bridge inspection of the structure. Using the example of a three-column bent for a girder bridge shown in Figure 3.2, moderate deterioration on the bottom of the pier beam under the curb line of the deck above should not be used to reduce the condition rating of the substructure. However, deterioration of the bottom of the pier beam between the columns should be considered in the condition rating. The bottom mat of reinforcing steel in between columns provides the positive moment capacity for the pier beam. The positive moment demand is greatest between the columns and minimal past the columns.

A history of bridge inspections for any particular structure can be used to track gross deterioration of the components over its life. For any inventory of bridges, rates of deterioration and estimates of service life for particular bridge and component types can be calculated from histories of condition ratings. This will be discussed later in the text in section 6.4 on service life. There are two major limitations to use condition ratings for programming bridge maintenance work. However, first, the component-level evaluation is too gross to capture rates of deterioration of finer levels of bridge members. For example, a deck consists not only of the structural members spanning between girders, but also of the expansion joints that typically deteriorate quicker than the rest of the deck. This has been addressed by the adaption of element-level inspection requirements by FHWA, which will be discussed shortly.

3.2.3 Appraisal Ratings

The second major limitation is that condition ratings are relative to as-built capacity, not to the current demands on the bridge. Truck sizes and traffic volumes have increased significantly over the past decades. A 70-year-old bridge in pristine condition still may not have sufficient structural capacity to carry modern large load-permitted trucks. It also may be too narrow to allow for the number of lanes required to carry current traffic without delay. Appraisal ratings are an evaluation of a bridge relative to current level of service requirements. There are seven items reported on the SI&A that require appraisal ratings.

- Structural evaluation;
- Deck geometry;
- Underclearances, vertical and horizontal;
- Waterway adequacy;
- Approach roadway alignment;
- Safety features; and
- Scour critical bridges.

Similar to the condition ratings, appraisal ratings are given on a 0–9 scale with the guidelines listed in Table 3.5.

The items of deck geometry, underclearances, and approach roadway alignment impact the number of lanes of traffic carried by bridge, the size of vehicles passable underneath it, and posted speeds on the bridge. Waterway adequacy describes the frequency of overtopping at the bridge and how it relates to current road design criteria. Safety features relates to level of protection provided by the existing roadside safety appurtenances (such as

Table 3.5 FHWA appraisal rating guidelines

Rating	Description
N	Not applicable
9	Superior to present desirable criteria
8	Equal to present desirable criteria
7	Better than present minimum criteria
6	Equal to present minimum criteria
5	Somewhat better than minimum adequacy to tolerate being left in place as is
4	Meets minimum tolerable limits to be left in place as is
3	Basically intolerable, requiring high priority of corrective action
2	Basically intolerable, requiring high priority of replacement
1	This value of rating code not used
0	Bridge closed

guard fence) to current standards. All of these items are of use when determining the scope of work for a bridge replacement or major reconstruction (e.g., if a bridge deck is to be replaced, it may make sense to widen the bridge to provide for additional lanes of traffic) but low rating for these items may not necessarily trigger bridge maintenance actions.

The two appraisal items that are more germane to bridge maintenance and preservation are the structural evaluation and scour critical ratings. To assign the structural evaluation rating, the lower of the superstructure and substructure condition ratings (or the culvert condition rating for a culvert) are compared to the value assigned in Table 3.1 in the *FHWA Recording and Coding Guide for the Structural Inventory and Appraisal of the Nation's Bridges* on the basis of the average daily traffic (ADT) carried by the bridge and its inventory load rating[1] for the HS truck. This rating not only reflects the condition of the primary structural components of a bridge, but also reflects its capacity relative to the demand from modern truck traffic loads (Table 3.6).

Table 3.6 Rating for structural evaluation for inventory load rating and ADT

Structural evaluation rating code	Inventory rating in metric tons for MS truck (HS truck equivalent in short tons)*		
	Average daily traffic		
	0–500	**501–5000**	**>5000 (or any interstate or expressway)**
9	>32.4 (HS20)	>32.4 (HS20)	>32.4 (HS20)
8	32.4 (HS20)	32.4 (HS20)	32.4 (HS20)
7	27.9 (HS17)	27.9 (HS17)	27.9 (HS17)
6	20.7 (HS13)	22.5 (HS14)	24.3 (HS15)
5	16.2 (HS10)	18.0 (HS11)	19.8 (HS12)
4	10.8 (HS7)	12.6 (HS8)	16.2 (HS10)
3	Inventory rating less than value in rating code of 4 and requiring corrective action.		
2	Inventory rating less than value in rating code of 4 and requiring replacement.		
0	Bridge closed due to structural condition.**		

* The inventory load rating values shown are from the December 1995 Coding Guide, which uses the International System of Units for measurements. The next version of the guide will use English units.

** As per the FHWA Coding Guide, a value of 1 is equivalent to a value of 0 for the structural evaluation rating.

[1] Load rating is discussed in Section 3.6. A load rating is an evaluation of the live load capacity of a bridge in terms of truck loadings. Inventory load ratings presume a stress level low enough for indefinite usage by the rating truck.

Table 3.7 Scour critical ratings below 4

Scour critical rating code	Description
4	Bridge foundations determined to be stable for calculated scour conditions; field review indicates action is required to protect exposed foundations from effects of additional erosion and corrosion.
3	Bridge is scour critical; bridge foundations determined to be unstable for calculated scour conditions, either: • Scour within limits of footing or piles; or • Scour below spread-footing base or pile tips.
2	Bridge is scour critical; field review indicates that extensive scour has occurred at bridge foundations. Immediate action is required to provide scour countermeasures.
1	Bridge is scour critical; field review indicates that failure of piers/abutments is imminent. Bridge is closed to traffic.
0	Bridge is scour critical. Bridge has failed and is closed to traffic.

The scour critical rating applies to those bridges with foundations that may be unstable in a high flow event resulting in scouring. The designation as a scour critical bridge is made as the result of an engineering assessment of the vulnerablity of the particular foundations of a bridge for either the observed or calculated potential scour at the bridge site. Appraisal ratings above 4 are for bridges determined to be stable for potential scour. Those below 4 may trigger maintenance actions to provide protection (such as placement of heavy stone rip-rap or scour mats) for the foundations (Table 3.7).

3.2.4 Deficiency and Sufficiency

Low condition or appraisal ratings will result in a bridge being categorized as deficient. A bridge that lacks geometric characteristics, load capacity, or waterway adequacy to support the expected level of service for the highway facility it carries may be considered functionally obsolete (FO). A bridge is classified as such if it receives an appraisal rating of 3 or less for deck geometry, underclearances or approach roadway alignment, or a rating of 3 for structure evaluation or waterway adequacy.

A bridge with sufficient deterioration or damage to one or more of its structural components, or which is inadequate in either its load capacity or waterway adequacy may be considered structurally deficient (SD). A bridge is classified SD if it receives a condition rating of 4 or less for either its deck, superstructure, substructure, or culvert components or if it receives an appraisal rating of 2 or less for structure evaluation or waterway adequacy. A bridge that meets criteria for both SD and FO (e.g., has a substructure condition rating of 4 and an underclearance appraisal rating of 3) is considered SD.

Deficient status is a flag to a reviewer that a bridge strongly needs to be considered for maintenance work or for replacement. However, this is a binary measure – a bridge is deficient or not. When programming maintenance and construction work for an inventory of bridges, a measure is needed that allows ranking for bridges according to need. Given a limited budget for maintenance and construction, prioritizing work also requires considering the importance of a particular bridge to operation of the highway network and of the level of service it provides.

To facilitate programming, FHWA created a performance measure known as the sufficiency rating (SR). This is a value, ranging from 0–100, where a value of 0 is considered entirely deficient and a value of 100 is considered entirely sufficient to remain in service. The four factors used to calculate the SR (SR = S1 + S2 + S3 − S4) are shown below:

- S1 = 55% maximum; based on structural adequacy and safety (i.e., superstructure, substructure, or culvert condition and load capacity).
- S2 = 30% maximum; deals with serviceability and functional obsolescence (items such as deck condition, structural evaluation, deck geometry, underclearances, waterway adequacy, approach road alignment).
- S3 = 15% maximum; concerns essentiality for public use (items such as detour length, ADT, and the Strategic Highway Corridor Network.
- S4 = 13% maximum; deals with special reductions based on detour length, traffic safety features, and structure type.

The factor, S4, concerns only possible reductions of the SR; however, the minimum value allowed for the SR is zero. Note that the official SR is calculated by the FHWA and reported back to the state or federal agency responsible for the bridge.

3.2.5 Critical Findings

In addition to condition and appraisal ratings, the NBIS requires bridge inspectors to note and to report any critical findings discovered during their investigations. Critical finding are defined as structural or safety-related deficiencies that require an immediate follow-up by either further inspection or action. Federal regulations leave it to each state or federal agency bridge owner to establish their own procedures to notify the FHWA of critical findings and to address them in a timely manner.

Structure-related deficiencies are defined in the *FHWA Bridge Inspector's Reference Manual* as those that interrupt the intended load path in the structure. With the diversion of load, surrounding elements may become overstressed or unstable, potentially leading to partial or total collapse of the structure. An example of this would be a shear crack in a girder, large enough to significantly reduce the capacity of the girder. This would result in load being shed to adjacent girder lines.

Safety-related deficiencies are those that may jeopardize the safety of motorists or pedestrians. A spall in the wearing surface of a deck deep enough to require covering with a steel plate would be an example of this.

By definition, critical findings require assessment for maintenance or construction actions. Typically the options available are to: temporarily repair and monitor the deficiency, to restrict loads (i.e., truck sizes) on the bridge, or to permanently repair the deficiency.

3.2.6 Element-Level Inspection

The NBIS has been successful in the goal of improving the safety of highway bridges in the United States for the traveling public. The number of deficient bridges on public roads in the United States is being steadily reduced. For the years for which bridge inspection data have been collected following the 1992 NBI coding guide (1992–2013) the percentage of deficient bridges has been reduced from 37.8% to 24.3% (Figure 3.3). The

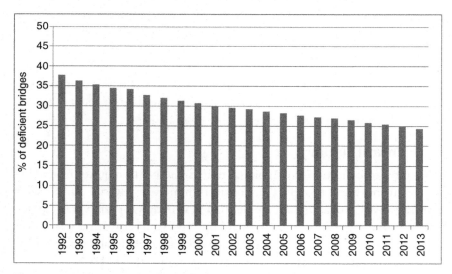

Figure 3.3 *Percentage of Deficient Bridges 1992–2013.*

system of biennial routine inspections, assorted special inspections, rating component conditions, assigning appraisal ratings, and addressing critical findings force bridge owner to address defects and dangerous conditions.

The recording of inspection data in inventory files systems has resulted in a collection of basic data on bridge conditions over the past several decades. This has provided owners the opportunity to map the deterioration rates of bridge components over time as they vary with material, construction type, and level of traffic. Data such as these form the basis for numeric models of bridge deterioration. The ability to model bridge deterioration makes prediction possible. Inspections that were initiated to catch defects and damage to bridges and protect the public have the potential to provide data that protects the public's investment.

As originally established, NBI condition ratings are ratings of components. Components are assemblies of individual members that comprise a distinct section of a bridge. For example, a slab, expansion joints, and railing together comprise the deck component in NBI condition ratings. These members each contribute to a common function, providing a roadway for highway traffic. But each of these members is made of different material and is subject to different loading demands; for example, the expansion joints are often constructed from steel and are subject to impact loading from each vehicle over a bridge, while railings may be constructed of reinforced concrete and are subject to loading only on rare occasions. A component condition rating is sufficient to alert to the presence of a dangerous deficiency, but it is too gross of a measure to capture the rate of deterioration of the individual members (Table 3.8).

In October 1989, the FHWA issued a final report for Demonstration Project 71, a review and introduction to Bridge Management Systems (BMS) [1]. A BMS consists of a database of bridge information and analysis tools to identify needs and prioritization for bridge maintenance, rehabilitation, and replacement. The initial work in various states to develop bridge management systems revealed that a finer resolution of bridge inspection data would be required. In the early 1990s, the FHWA sponsored the development of the Pontis Bridge Management System. This BMS was based on bridge elements rather than the three NBIS structural components for span bridges.

Bridge elements are individual members comprised of basic shapes and materials performing the same function. While the superstructure of a bridge with a steel girder and floorbeam system is only one component for NBIS component condition ratings, there are separate structural elements of steel girders, floorbeams, stringers, and bearings for an element-level inspection. To provide uniformity in element definitions among states, after

Table 3.8 NBI component condition ratings for concrete deck per KDOT Inspection Manual

Rating	Description	Criteria
9	New Structure – not open to traffic	
8	Good Condition – no repairs needed	• No significant spalling, scaling, delamination, or map cracking • No water saturation • Minor transverse cracking (up to 0.5 mm), very isolated
7	Generally Good Condition – repairable by AREA	• Deck cracks with or without efflorescence (up to 1.00 mm cracks) can be sealed • Light scaling (0.25 in. depth or less) • Visible tire wear in the wheel paths • 10% or less of the deck is stained, deteriorated, and/or hollow planed • Minor surface spalls • Light map cracking (up to 0.5 mm)
6	Fair Condition – repairable by district	• 2% or less of the deck is spalled exposing reinforcing steel • Medium scaling (0.25–0.5 in. in depth) • 10–20% of the deck is stained, deteriorating, and/or hollow planed. Note: debonded overlays • Deterioration of deck edges or outlets • Excessive number of open cracks (excessive being at 5 ft. intervals or less over the entire deck) (1.00–1.50 mm) • Map cracking moderate (up to 1.00 mm) may soon lead to spalling • All concrete decks with a bituminous wearing surface will be rated no higher than 6
5	Generally Fair Condition – would require contract repair	• 2–5% of the deck is spalled exposing reinforcing steel • Excessive cracking resulting in spalling (>1.50 mm) • Heavy scaling (0.5–1 in. in depth) • 20–40% of the deck is stained, deteriorating, and/or hollow planed • Disintegration of deck edges or outlets

Table 3.8 NBI component condition ratings for concrete deck per KDOT Inspection Manual *(cont.)*

Rating	Description	Criteria
4	Poor Condition – section loss, deterioration, or spalling in the deck or slab are in an advanced state that warrants posting structure more than 20 tons	• More than 5% of the deck is spalled exposing reinforcing steel • 40–60% of the deck is stained, deteriorating, and/or hollow planed
3	Serious Condition – section loss, deterioration, or spalling in the deck or slab are in an advanced state that warrants posting structure from 10 tons to20 tons	• More than 60% of the deck is stained and/or deteriorating • Many full-depth failures with deck plates or in need of plates placed • Severe or critical signs of structural distress are visible on bridges where the deck is integral with the superstructure (i.e., deck sagging).
2	Critical Condition – facility should be closed until the indicated repair is completed	
1	Critical Condition – facility is closed. Study should determine the feasibility for repair, replacement, or removal	
0	Critical Condition – facility is closed and is beyond repair	

the initial release of Pontis in 1991 AASHTO began work on establishing definitions of elements and condition states that would be uniform between states [2]. The result was the publishing of the *AASHTO Guide for Commonly Recognized (CoRe) Structural Elements*.

Generally, different CoRe elements have distinct functions, deterioration rates, and maintenance requirements. CoRe elements are quantified in terms of length, area, or each (for enumerate elements such as bearings). Terms of quantification are chosen in regard to utility for inspections and for scoping maintenance work. For example, a bridge inspector can readily measure girders by length in the field, but cannot determine the weight of a particular section of girder.

The condition states of CoRe elements are reported on a scale of 3–5 different defined states. Structural elements tend to follow a five-point scale that according to Thompson [2], was modeled on a common "scale that reflects the most common processes of deterioration and the effect of deterioration on serviceability." The pattern is: protected, exposed, attack, damaged, and failed. For a concrete bridge deck, deterioration and damage typically

Table 3.9 Deck condition states in Pontis from the KDOT Inspection Manual

Condition state	Description
1	• Repaired areas and/or spalls/delaminations can exist in the deck surface. • The combined distressed area is <2% of the deck area.
2	• Repaired areas and/or spalls/delaminations exist in the deck surface. • The combined distressed area is ≥2 to ≤10% of the deck area.
3	• Repaired areas and/or spalls/delaminations exist in the deck surface. • The combined area of distress is >10 to ≤25% of the total deck area.
4	• Repaired areas and/or spalls/delaminations exist in the deck surface. • The combined area of distress is >25 to <50% of the total deck area.
5	• Repaired areas and/or spalls/delaminations exist. • The combined area of distress is ≥50% of the total deck area

take the form of cracks, spalls, and delaminations. A pristine concrete deck surface might be presumed to be shielding the reinforcing steel in the slab from corrosion. As it cracks, however, water and road salts may penetrate to the reinforcing steel and initiate corrosion. Condition states for the deck elements are related to the relative quantity of distressed (cracked, spalled, or delaminated) surface area (Table 3.9).

A useful feature of element-level inspection is the ability to quantify the portion of an element that may be damaged or distressed. Observe on a section of a sample bridge inspection form from KDOT below that there are 300 ft. of concrete bridge rail in condition state 1 and 45 ft. in condition state 2. Information of this detail can be extremely useful for those scoping bridge maintenance work (Figure 3.4).

Note that the concrete deck element is all in a single condition state, state 2. Also note that there is no way to distinguish the type or mode of deterioration. A bridge deck, in particular, may exhibit multiple modes of distress. For example, there may be cracking in the positive moment region over the pier in a deck composite with steel girders while at the curb lines there may be spalls caused by the separate issue of poor drainage.

As the use of CoRe elements and Pontis has proliferated, the potential for improvements became apparent. A number of these improvements have

District: 1	Area: 4	Sub-Area: 1	Route: U 24	County: 044	Ref Pt: 374.23	Serial No.: 0003

Unit: 1 RDGH - 4   Wide: None   Spans: 1@ 50 , 1@ 70 , 1@ 50

____ DECK:

DkRate: 7

Deck_HI: 75

Crit.Notation: None

Deck Material:	Reinforced Concrete	Deck Thickness:	7.3 in.
Wearing Surface:	Silica Fume	WS Thickness:	1.5 in.
Median:	No Median	Median Width:	0.0 ft.
Railing:	Conc Parapet/Alum	Deck Width:	33.3 ft.
Curb/Sidewalk Width	Left: 0.0 ft. Right: 0.0 ft.		
Near Exp Device:	N/A		
Far Exp Device:	N/A	Deck Drain Sys: Metal Slots&Slope Drains	

Delams: ____ 8 %

TOP:	Deter	8 %
	Spalls	1 %
BTM:	Deterioration	5 %
	Spalls	0 %

PONTIS Element Description	Env.	Qty.	Units	Cond 1	Today	Cond 2	Today	Cond 3	Today	Cond 4	Today	Cond 5	Today
22 P Conc Deck/Rigid Ov	3	4830	(sf)	0	____	4,830	____	0	____	0	____	0	____
321 R/Conc Approach Slab	3	2	(ea)	2	____	0	____	0	____	0	____	0	____
331 Conc Bridge Railing	3	345	(lf)	300	____	45	____	0	____	0	____	0	____
358 Deck Cracking SmFlag	1	1	(ea)	0	____	1	____	0	____	0	____	0	____
359 Soffit Smart Flag	1	1	(ea)	0	____	1	____	0	____	0	____	0	____

____SUPER:

SupRate: 7

Super_HI: 98.5

Crit. Notation: None

Hinge Type:	Not Hinged	Bearing Type:	N/A
Paint Condition: N/A		Change to: ____	
Last Painted:	Type: N/A	Steel: N/A	

PONTIS Element Description	Env.	Qty.	Units	Cond 1	Today	Cond 2	Today	Cond 3	Today	Cond 4	Today	Cond 5	Today
110 R/Conc Open Girder	2	680	(lf)	650	____	30	____	0	____	0	____	0	____

____SUB:

SubRate: 7

Sub_HI: 98.8

Crit. Notation: None

Abutment Type Near:	Conc U-Type On	on	Steel H-Pile
Abutment Type Far:	Cap on "Integral"	on	Steel H-Pile
Pier Type:	Web Frames	on	Steel H-Pile
Berm Protection:	Earth Berms		

PONTIS Element Description	Env.	Qty.	Units	Cond 1	Today	Cond 2	Today	Cond 3	Today	Cond 4	Today	Cond 5	Today
205 R/Conc Column	2	8	(ea)	8	____	0	____	0	____	0	____	0	____
215 R/Conc Abutment	3	67	(lf)	60	____	7	____	0	____	0	____	0	____
361 Scour Smart Flag	1	1	(ea)	1	____	0	____	0	____	0	____	0	____

Figure 3.4 *Sheet From Typical KDOT Bridge Inspection.*

been codified in the next generation of element-level inspection in the *AASHTO Guide Manual for Bridge Element Inspection* first published in 2011[2] (Table 3.10). Some of the improvements include:

- Development of separate bridge elements for protective elements, such as, distinct elements for the slab and for any overlay;

[2] This manual was superseded in 2013 with the publishing of the AASHTO Manual for Bridge Element Inspection, First Edition.

Table 3.10 Condition states for reinforced concrete decks from AASHTO Manual for Bridge Element Inspection

	Condition states			
	1	2	3	4
Defects	Good	Fair	Poor	Severe
Delamination/spall/patched area (1080)	None	Delaminated spall 1 in. or less deep or 6 in. or less in diameter. Patched area that is sound.	Spall greater than 1 in. deep or greater than 6 in. diameter. Patched area that is unsound or showing distress. Does not warrant structural review.	The condition warrants a structural review to determine the effect on strength or serviceability of the element or bridge; or a structural review has been completed and the defects impact strength or serviceability of the element or bridge.
Exposed rebar (1090)	None	Present without measurable section loss.	Present with measurable section loss but does not warrant structural review.	
Efflorescence/rust staining (1120)	None	Surface white without build-up or leaching without rust staining	Heavy build-up with rust staining.	
Cracking (RC and other) (1130)	Width less than 0.012 in. or spacing greater than 3.0 ft.	Width 0.012–0.05 in. or spacing of 1.0–3.0 ft.	Width greater than 0.05 in. or spacing of less than 1 ft.	
Abrasion/wear (PSC/RC) (1190)	No abrasion or wearing	Abrasion or wearing has exposed coarse aggregate but the aggregate remains secure in the concrete.	Coarse aggregate is loose or has popped out of the concrete matrix due to abrasion or wear.	
Damage (7000)	Not applicable	The element has impact damage. The specific damage caused by the impact has been captured in condition state 2 under the appropriate material defect entry.	The element has impact damage. The specific damage caused by the impact has been captured in condition state 3 under the appropriate material defect entry.	The element has impact damage. The specific damage caused by the impact has been captured in condition state 4 under the appropriate material defect entry.

- Allowing decks to be allotted into multiple condition states with type of defect noted;
- Distinguishing edge of deck, which is typically subject to greater levels of deterioration than the interior; and
- Establishing National Bridge Elements and Bridge Management Elements, standardized elements that are indicative of the structural condition and of maintenance needs, respectively. Agencies are also free to define Agency Defined Elements for their own use.

FHWA has mandated reporting of element-level data per the *AASHTO Manual for Bridge Element Inspection* concurrent with NBI condition ratings for all states and agencies starting October 2014.

3.3 BRIDGE INSPECTIONS IN CANADA, WESTERN EUROPE AND SOUTH AFRICA

3.3.1 Canada

Like the United States, there are three levels of government that have jurisdiction over bridges in Canada. In Canada, they are the Federal, Provincial or Territorial, and Municipal. According to a 2012 thesis by Khanzada, there are over 74,000 bridges in Canada; 43,003 owned at the Municipal level, 30,894 at the Provincial or Territorial level, and 419 at the Federal level [3]. Like the United States, most of these structures were built in the years soon after World War II. Over 40% of these bridges are over 50 years old [4]. Unlike the United States, there is no federal specification concerning the inspection and management of bridges. The Federal government is directly responsible for less than 1% of bridges on public roads in Canada. The provincial/territorial and the municipal governments are responsible for almost all highway bridges in Canada. In most of the provinces and territories, there is no central database inclusive of municipal bridges in the province or territory, nor is there centralized oversight or regulations for inspection. As a result, for bridges on Canadian public roads, there are multiple methodologies for inspection and for recording conditions.

Each province and territory has adopted one of four manuals to govern bridge inspections on their own system. British Columbia has its own *Bridge Inspection Manuals – Books 1, 2, and 3* dated 1994. Alberta's *Bridge Inspection and Maintenance Manual* dated 2005 is used and the Yukon and the Northwest Territories. The *Ontario Structure Inspection Manual* is the most widely used, adopted by five other provinces (Manitoba, New Brunswick, Nova Scotia, Prince Edward Islands, and Saskatchewan). Quebec has its own French language manual *Manuel d'Inspection des Structures* dated 2012.

In British Columbia, bridge components are rated on a scale of 1 (excellent) to 5 (very poor). Bridges receive a visual inspection annually by province inspectors. A detailed inspection with equipment to provide up-close access is suggested (not required) to be conducted every 5 years. Inspectors are qualified through training by the province.

The Alberta *Bridge Inspection and Maintenance Manual* uses ratings of 1 (immediate action) to 9 (very good) for bridge components similar to the FHWA NBIS component condition ratings. The current condition of an element is related to the original condition, not to current design standards. Inspections are categorized into level 1, a general detailed visual inspection by a minimum of a class B inspector and level 2, an in-depth inspection by a class A inspector. Level 2 inspections include more detailed measurements and close access. They are required for some nonstandard bridges or when particular deficiencies are found. Certification as a level 1 inspector requires a high-school diploma and completion of a training course and fieldwork by the province. Certification as a level 2 inspector requires an engineering degree and 2 years of bridge experience or an engineering technology technical diploma and 3 years of bridge experience in addition to the training course and fieldwork.

In Alberta, bridges on major provincial highways are required at a maximum interval of 21 months, all bridges on secondary highways and major bridges on local roads at 39 months; and standard bridges on local roads at 57 months. In the Yukon, level 1 inspections are conducted every 2 years, while in the Northwest Territories the interval is 3 years.

The Ontario Structure Inspection Manual uses four condition states (excellent, good, fair, and poor) to rate bridge elements, similar to element-level inspection implemented in the United States. Similarly, the condition state may apply only to a portion of an element; for example, a bridge deck may have 70% of its area in excellent condition, 25% in good condition, and 5% in poor condition.

In Ontario, a detailed visual inspection is required every 2 years and must be done by or under the direction of a PE. The inspectors must complete a training course and have had a minimum of 3 years' experience with structure maintenance practices.

In Saskatchewan, major bridges are to be inspected by a two-person team led by either a PE or by an engineering technologist with 5 years' inspection experience. The lead must also have completed a bridge inspection course. Detailed visual inspections are conducted biennially. For bridges with concrete decks lacking a waterproofing membrane, a deck survey is conducted every 4 years.

The other four provinces (Manitoba, New Brunswick, Prince Edward Islands, and Saskatchewan) using the Ontario manual have inspectors complete training either in-house or utilize courses associated either with the *Ontario Structures Inspection Manual* or with the FHWA. The inspection intervals vary with each province. New Brunswick conducts regular bridge inspections every 2 years. Prince Edward Islands has bridge inspectors conduct walk-around inspections every 3 years with intervals for regular inspections determined by the findings for each bridge. Manitoba varies the interval for detailed visual inspections by the road classification, with major bridges on Provincial Trunk Highways inspected every 2 years, major bridges on provincial and main market roads inspected every 4 years, and minor bridges on provincial, main market, access, and service roads inspected every 6 years. Nova Scotia varies inspection intervals based on the overall condition of the bridge from the last inspection report. Bridges in good condition are inspected at 4–5-year intervals, those in fair condition at 2–3 years, and bridges in poor condition are inspected annually.

In 2008, Quebec implemented a new bridge management system, the Système de Gestion des Structures. In support of the system it adopted element-level inspections, using a scale of A (excellent) to D (poor) applied to the portion of the element in that particular condition. Inspections are biennial. Both engineer and engineering technician inspectors are trained in-house.

One feature common throughout most of the provinces and territories is a requirement for maintenance inspections, that is, inspection by the field personnel responsible for day-to-day maintenance, on at least an annual basis. This is true whether the road facility is being maintained by employees of the province or territory, or whether maintenance had been privatized and contracted.

3.3.2 United Kingdom

In 2006, the FHWA conducted a scanning tour of bridge inspection practices in Europe and South Africa [5]. The results of the tour initiated further work including an NCHRP report on bridge inspections practices in the United States and abroad. [6]. The discussion for practices in the United Kingdom and other countries is based on the results from the 2006 scanning tour and supplemental materials.

The United Kingdom Highways Agency is responsible for 9,400 km (5,840 miles) of trunk roads across the nation, including approximately

10,000 bridges. A structure is defined as a bridge if it has a span of 1.8 m (5.9 ft.) or greater. Each bridge is subject to a general inspection at a maximum interval of 2 years and to a principal inspection at a maximum interval of 6 years. A general inspection is a visual inspection of all members of the bridge that can be seen without equipment for access. A principal inspection is a close examination to ascertain the condition of all parts of the structure. The condition of the members is recorded during principal inspections in terms the severity of the defect, on a scale of 1–5, and the extent of the defect, on a scale of A–E.

At the time of the FHWA scanning tour, there were no formal regulations regarding the certification of bridge inspectors. In 2012, the UK Bridges Board Group, a group of local and national governmental organizations tasked with developing standards and best practices for constructing and operating civil engineering structures, published a draft copy of bridge inspectors training standards [7]. If adopted, the guidelines call for on-the-job training of bridge inspectors, with two levels of certification: inspector (minimum of 2 years' experience) and senior inspector (minimum of 5 years' experience). An inspector could conduct principal inspections on simple bridge types. A senior inspector would be required for more complex bridges.

3.3.3 South Africa

South Africa has 62,000 km (38,500 miles) of paved roads, 7,200 km of which are trunk roads. There are approximately 2100 bridges (as defined by a span of at least 6 m (19.7 ft.)) on the trunk roads. As of 2006, 58% of the trunk road network's miles were over 21 years old. Other than maintenance inspections, the routine bridge inspection is a principal inspection, conducted on a 3–5-year cycle by a PE.

Bridge inspection reporting is in terms of defects rather than conditions. Defects are assigned ratings for degree, extent, and relevancy. The rating values range from 0–4 (Table 3.11).

3.3.4 France

In 2008 there were approximately 230,000 highway bridges in France. Departments (states) and local entities are responsible for 85% of these structures. Of the remainder, less than 15,000 are maintained directly by the French National Road Directorate. The others are roads conceded to maintenance by private companies. Note, a minimum span for a structure to be considered a bridge in France is 2 m (6.6 ft.).

Table 3.11 DER ratings for inspections in South Africa

Category	0	1	Rating values 2	3	4
D – Degree of defect	Severity of defect.	None	Minor	Fair	Poor	Severe
E – Extent of defect	Prevalence of defect within the element.		Local	>Local	<General	General
R –Relevancy of defect	Impact of the defect on structural integrity and/or user safety.		Minimum	Moderate	Major	Critical
U – Urgency of defect	Recommended time for repair.	Monitor only	Routine	<5 years	<2 years	ASAP

Inspection procedures are codified in the *Instruction Technique pour la Surveillance et l'Entretien des Ouvrages d'Art* (ITSEOA). At the time of the scanning tour, condition ratings were reported on a scale of 1 (good) to 3 (damaged) with subdivisions for urgency of maintenance. In 2010, a new version of the ITSEOA was published with a migration to an element-level bridge inspection methodology.

There are three levels of certification for bridge inspectors: team leader (chargé d'étude), inspector, and inspection agent. Inspection agents must have the equivalent of a high-school diploma and are certified by the laboratory director where employed. Inspectors must have the equivalent of a college degree, 2 years of experience, have completed training recognized by the certifying board of the Network of Roads and Bridges Laboratories and have passed examination and review by the board. Team Leaders must have a degree in civil engineering, 5 years of experience and passed training, examination, and review by the board.

Cursory inspections are conducted annually to look for obvious changes in condition. Visual inspections conducted by Inspection Agents are conducted every 3 years to rate the condition of the bridge in Image de la Qualite d'Art standards. Detailed inspections are conducted nominally every 6 years by an inspector. Robust bridges may be placed on a 9-year cycle while bridges in poorer conditions may be placed on a 3-year, or even annual, cycle. The detailed inspection is conducted within arm's reach of every component.

3.3.5 Germany

The German road network is 626,000 km (389,000 miles) long, with 53,000 km of federal roads that are administered by the Bundesministerium für Verkehr, Bau und Wohnungswesen (BMVBW) (Federal Ministry of Transport, Building, and Housing). In 2008 there were 120,000 bridges on all roads with 37,000 of them on federal roads. Standards set by BMVBW are mandatory for only federal bridges.

The two levels of routine inspection by bridge inspectors in Germany are major and minor tests. Major tests are arm's reach inspections of every component of the bridge and are on 6-year intervals. Minor tests are conducted 3 years after each major test, typically without access equipment with the purpose of reviewing defects noted on the last major test. Road maintenance crews are required to visit each structure every 3 months as a cursory safety inspection. Bridges inspectors are required to have a civil engineering degree, 5 years of bridge engineering experience and to complete an in-house inspection course.

Table 3.12 Damage assessments for structural elements in Germany

Assessment of damage	Structural stability	Traffic safety	Durability
0	No effect	No effect	No effect
1	Negatively affects element, no effect on structure.	Affects only slightly, traffic safety is given. Repair with regular maintenance.	Negatively affects element, no long-term effect on structure.
2	Negatively affects element, little effect on structure.	Affects only slightly, traffic safety is given. Repair or place warning signs.	Negatively affects element and, in the long term, the structure. Expansion of damage may occur.
3	Negatively affects element, effect on structure exceeds permissible tolerances.	Affects traffic safety. Repair or place warning signs.	Negatively affects element and, in the medium term, the structure. Expansion of damage is expected.
4	Stability of element and structure no longer exists.	Traffic safety is no longer given. Restrict use immediately.	Durability of element and of structure is no longer a given. Immediate repairs or restrictions required.

Structural elements are assessed for damages on a 0–4 scale each for: structural stability, traffic safety, and durability (see Table 3.12) [8]. The inspection program used then calculates a composite condition index for each element.

3.3.6 Finland

The Finnish Road Administration is responsible for 11,000 of the approximately 18,000 highway bridges in Finland. Of the rest, 6000 are private bridges and 1800 are municipal bridges. In Finland, a bridge has a minimum span of 2 m (6.6 ft.).

General inspections are made every 5 years for most bridges, every 8 years for large bridges, by certified bridge inspectors. All components are inspected at arm's length. Defects are given ratings in four categories: weight

Table 3.13 Defect ratings in Finland

Structural part							
Substructure	0.70	0 – New or like new	1	11 – Repair during the next 2 years	10	1 – Mild	1
Edge beam	0.20	1 – Good	2			2 – Moderate	2
Superstructure	1.00	2 – Satisfactory	4	12 – Repair during the next 4 years	5	3 – Serious	4
Overlay	0.30	3 – Poor	7			4 – Very Serious	7
Other surface structures	0.50	4 – Very Poor	11	13 – Repair in the future	1		
Railings	0.40						
Expansion joints	0.20						
Other equipment	0.20						
Bridge site	0.30						

(importance in the load path), condition of the structural element (apart from the defect), urgency of the repair, and damage class (severity of the defect). The ratings are used to calculate a repair index, KTI, for the bridge.

$$KTI = \max(Wt_i \times C_i \times U_i \times D_i) + k\sum(Wt_j \times C_j \times U_j \times D_j)$$

Equation 3.1 Repair Index Formula for Bridges in Finland.

Values for the variables are given in Table 3.13. Note that the subscript, i, indicates the worst defect; while the other defects are indicated by the subscript, j. The variable, k, is a weighting factor for summation, with a default value of 0.2.

Inspector certification requires completion of a 4-day training course and 2 days of fieldwork followed by testing. Annual training is required, during which each inspector performs general inspection on two bridges. Large deviations from the expected inspection results may result in loss of certification. To qualify for certification as an inspector for general inspections, the equivalent of a high-school diploma is required.

Finland maintains a group of approximately 125 reference bridges. These bridges are subject to basic inspections by engineers who are certified bridge inspectors. Basic inspections include material tests, including those for chloride content and carbonation depth. The results are used to improve knowledge of material performance and bridge service life.

3.3.7 Observations

The countries examined face a situation of managing highway bridges similar to that faced by the United States, the construction of most bridges is similar to that in the United States, and most countries had a significant expansion of their highway systems in the decades right after World War II.

The countries did differ among each other in two important aspects in managing their bridge inventories. They differed in approaches to inspector requirements and certification, some requiring engineers for all, others only for advanced inspections. They also differed in their approaches to condition ratings. Many concentrated on reporting defects rather than overall conditions. Others countries were migrating to element-level inspections.

The studied countries differed from the United States in two particular aspects. Most other countries did not have national-level laws regulating inspections, nor did the larger countries have a national database similar to the United States. Many of the countries had longer periods between bridge inspections made by certified inspectors than the United States and they made use of maintenance personnel to observe changes in condition on an annual basis or less. Note in the United States, it is typical for local maintenance personnel to monitor bridges, along with culverts, signs, lights, and other items in the roadside environment; however, the results are not typically recorded or reported in a formal process.

3.4 RELIABILITY-BASED BRIDGE INSPECTION

3.4.1 Risk-Based Assessment

A few years after the scanning tour, the NCHRP project 12-82 was initiated with the goal of developing a risk-based methodology to assign intervals to bridge inspection in the United States. European practices had shown that longer intervals between routine bridge inspections were feasible, if supplemented with maintenance inspections. Current practice in the United States does allow for increasing bridge safety inspection intervals from 24–48 months with the approval of the FHWA. As of 2014, only 15 states utilized this option. In practice this is typically reserved for small bridge-sized culvert structures and bridges on routes with low traffic counts, often in remote areas.

The project resulted in NCHRP Report 782, *Proposed Guidelines for Reliability-Based Bridge Inspection Practices* [9]. These guidelines allow a bridge owner to conduct a risk-based assessment of its inventory of bridges and to adjust the inspection interval for particular bridges down to 12 months or out to 72 months, depending on the type of construction and on the condition of the bridge. The stated intent was to allow for a more efficient allocation of inspection resources and to concentrate on areas of greater concern for public safety.

Three definitions key to the methodology are defined in the report as:

- Risk – the product of the probability of the occurrence of an event (in this case, failure of a bridge element) and the consequence of the occurrence of the event.
- Failure of a bridge element – the state when an element is no longer performing its intended function of safely and reliably carrying normal loads and maintaining serviceability.
- Reliability – the probability that failure will not occur during a given time period.

Given this, the reliability of a bridge element is the likelihood that it will continue function as intended for a given time period; that is, it will not suffer sufficient deterioration or damage to impede functioning. That likelihood is a function of the initial design and construction of individual elements, of the loading and environment to which it is exposed, and to the condition of the element at the beginning of the time period.

For any particular bridge or group of bridges, the first step in an assessment to adjust the inspection interval is to assemble a risk–assessment panel. This is a panel of experts in bridge inspection, maintenance, materials, and program management who are familiar with the inventory at hand. In their assessment, they will:

1. Decide which elements, or components, will be the focus of review. This depends on bridge inspection practices and on which elements are determined to be most critical to safety and serviceability.
2. Decide what metric constitutes failure, that is, at what state quantified in the inspection is a bridge element considered failed.
3. Determine the critical damage modes.
4. Determine which attributes of an element effect its condition for the critical damage modes. Note, one major attribute indicative of the performance of an element for a given time period is the condition of that element at the start of the period.
5. Establish occurrence factors, that is, estimates of the likelihood that the damage mode will result in a failed state of the element during the evaluation period of 72 months, for the range of attributes that an element may possess.
6. Concurrently and independently of establishing occurrence factors, establish consequence factors for failure from the damage mode of interest for the element of interest.

7. Given defined occurrence and consequence factors for a damage mode for a bridge element, develop a risk matrix relating the factors to proposed inspection interval.

8. Using the current attributes of a bridge element, the proposed period to the next bridge inspection is determined. For a particular bridge, several damage modes for several elements may be considered to find the controlling one during any particular evaluation.

Given an inventory of bridges and a history of bridge inspections, these results may be verified by backcasting. Backcasting involves reviewing the inspection history of particular bridges and determining if the inspection intervals determined by the methodology would have captured changes in bridge condition appropriately.

3.4.2 Implementation in Indiana

In October 2014, a risk-assessment panel workshop was held in Indianapolis, Indiana [10]. During this review with the Indiana Department of Transportation, decks, steel superstructures, and prestressed concrete superstructures were identified as components of concern for the process. For all of the components, an NBI condition rating of 3 was determined to be the failed condition.

Following the risk-assessment process for decks, the primary damage mode of concern was determined to be corrosion of the reinforcing steel. Other damage modes were discussed: cracking, rubblization, rutting, and debonding. Cracking was determined to be interrelated to corrosion. Debonding was found to be related to overlays rather than the deck elements, while rubblization and rutting were found to not be significant damage modes.

The panel then determined which attributes concerning the deck of a bridge would be relevant to assessing its reliability and durability. The most relevant were determined to be the current deck condition, its maintenance cycle and its exposure environment. Seven other attributes were identified and a matrix was created which assigned points for the condition of each of the attributes for each deck to be reviewed. The most relevant attributes were determined to have the most severe effect on the reliability and durability of the deck and were allotted to most possible points. The more points received by a reviewed deck the greater the occurrence factor for the particular deck.

Four screening attributes were also identified, each of which would exclude the bridge deck from review using the matrix shown in Table 3.14. Bridges with noncomposite superstructures were considered to have increased

Table 3.14 Attributes for deck corrosion damage modes

Attributes	High	Medium	Low	Remote	Maximum score
Current deck condition	Condition rating of 5	Condition rating of 6	Condition rating of 7 or more		20
Maintenance cycle	No maintenance			Washing/sealing	20
Exposure environment	Northern districts	Central districts	Southern districts		20
Efflorescence/leaching	Efflorescence with rust stains	Moderate efflorescence	Minor efflorescence without rust	No efflorescence	15
Concrete cover	<1.5 in.	1.5–2.5 in.	≥2.5 in.		15
Reinforcement type	Not epoxy coated			Epoxy coated	15
Deck drainage	Ponding/ineffective drainage			Effective drainage	15
Presence and type of overlay	Bituminous without membrane			No overlay or latex-modified concrete overlay	15
ADT and functional classification	>2500 –Interstate			<100 – Rural	15
Presence of repairs	Yes			No	10

Source: From an FHWA document (public use allowed).

reliability concerns and so would not be considered for longer inspection intervals. Bridges without a concrete deck would obviously not be eligible for review by these criteria. Bridges with a known construction error were also excluded from review for an extended inspection interval. Bridges with a deck with an NBI condition rating of 4 or less were determined to need an inspection interval of 24 months or less.

For this damage mode, the consequence factor for decks was determined to be a function of the site conditions at each individual bridge reviewed. The consequence factor for decks on bridges over roadways and multiuse paths was determined to be high due to the possibility of debris falling onto traffic below. Bridges over nonnavigable waterways and railways were determined to have medium to low consequence factors, dependent on the facility carried and the volume of traffic.

Given the consequence and occurrence factors determined for the deck of each bridge eligible for review, a proposed bridge inspection interval was determined based on the risk matrix. This was also done for the superstructure and any components or conditions reviewed for each bridge (Table 3.15). The most conservative of the proposed inspection intervals for each bridge was assumed to control.

Among the bridges reviewed in this workshop, most would be eligible for extended inspection intervals of 48–72 months. Based on families of bridges, 20% of state bridge inventory would be eligible for extended inspection intervals based on the assessment done in the 2014 workshop, with eligibility extending to possibly 60% of the inventory with further assessment.

Extending the inspection interval for a large portion, or even the majority of bridges in any owner's inventory offers considerable opportunity for savings and for reallocating inspection resources to bridges in more critical condition. However, it should be recognized that regular inspection intervals across an inventory also provides the owner with data concerning typical bridge deterioration rates in a population.

Table 3.15 Risk matrix of inspection intervals for IDOT

Occurrence factor	High	48 months	24 months	24 months	12 months
	Moderate	48 months	48 months	24 months	24 months
	Low	72 months	72 months	48 months	24 months
	Remote	96 months	72 months	48 months	48 months
		Low	Moderate	High	Severe
		Consequence factor			

Source: Adapted from Reising et al. [10]

3.5 INSPECTION TECHNIQUES AND TECHNOLOGIES

3.5.1 Visual Inspections and Sounding

The vast majority of bridge inspection is conducted with the unaided eye. For up-close and more in-depth inspections, visual observation is typically supplemented with physical sounding of the bridge members. Physical sounding of the members involves either tapping with a hammer, or dragging a metal chain across the surface of a member to listen for changes in tone resulting from variations in the condition of substrate material. A chain dragged across a concrete bridge deck will produce a high-pitched sound that drops in tone when the chain crosses an area which is voided or delaminated under the surface.

Visual observation will detect cracks, gouges, spalls, or other discontinuities in the surface and changes in color or texture that may be indicative of the presence of corrosion products. However, to be detected these defects must be present at the surface and of sufficient size to be perceived by a person with average visual acuity whose is standing within a few feet. Often, success in addressing defects by maintenance actions depends on performing the work while the scale of the defect is still relatively small. To detect defects that may not be apparent by unaided visual observation alone, inspectors may avail themselves of a number of test procedures specific to the material being tested and the defect of interest.

Test procedures may also have several other important benefits for the inspections process. First, measurements by testing are typically more objective and more readily quantified than results that are derived from the direct visual or audio (in the case of physical sounding) perceptions of an inspector. Second, the portions of a bridge most subject to wear and damage are often those directly exposed to automobile traffic. There are test methods that may be conducted either at a distance or from a moving vehicle that may either eliminate, or at lease minimize, the exposure of inspectors to danger from traffic.

Regarding the first point mentioned previously, in 2001 the FHWA published a report, *"Reliability of Visual Inspection for Highway Bridges"*. Researchers tested 49 inspectors representing 25 states conducting NBI component inspections on a series of test bridges. It was found that 32% of the ratings assigned varied two or more points from the average assigned value. Recommendations from the report to promote consistency in assigning ratings included testing of inspectors' visual acuity and color vision, implementing quality assurance procedures in each state's program, and providing inspectors with prompts.

3.5.2 Nondestructive Testing – Concrete

During the construction and fabrication processes, the materials used to form bridge members are sampled and subject to destructive testing to determine the properties of the materials. For example, samples of the concrete used are cast into cylinders that are broken later in a hydraulic press in order to determine the strength of the concrete placed. However, once a bridge is in service, taking samples out of the bridge members for testing is obviously impractical (though useful on specific occasions which will be discussed later). Testing methods for materials that do not involve removing samples are known as NDT or, alternatively, as methods of NDE. NDT methods and techniques are continually being updated as technology improves. This text is not indented to provide a comprehensive guide to conducting NDT methods. The more commonly used tests are discussed briefly later in the chapter, specifically in terms of providing information for scoping maintenance actions.

The most commonly used material in modern highway bridge construction is reinforced concrete. The presence of reinforcing steel in the concrete matrix leaves it susceptible to delamination and spalling due to corrosion of the reinforcing steel. The concrete matrix itself may degrade due to deficiencies incipient in the original construction or due to chemical attack during the life of the bridge.

Nondestructive tests used to detect delaminations under the concrete surface and to detect defects in concrete matrix include:

- Impact echo – Originally developed at the National Institute of Standards and Technology in the 1980s [11], the stress waves resulting from the action of an impacting device are measured. The interfaces between differing materials and layers of material result in different responses, allowing the detection of delaminations and voids.
- Ultrasonic techniques – In ultrasonic methods, an electromechanical transducer generates ultrasonic stress waves. A receiver measures the amplitude and velocity of the waves through the material that is affected by the density and elastic properties of the materials encountered. The location of reinforcing in and of cracks through concrete can be detected. The strength of the concrete can be estimated if calibrated on samples tested by destructive testing.
- Ground penetrating radar – High-frequency electromagnetic waves are propagated through concrete. Measurements of the responding signal can be used to map the location of reinforcing in the member and determine the thickness of the member. A comparison of NDT methods for detecting

concrete bridge deck deterioration [12] found that ground-penetrating radar had trouble detecting delaminations without the presence of accompanying moisture or chlorides in the defect. However, high grades were given for the speed at which a deck survey could be conducted with radar.

- Infrared thermography – Infrared cameras are used to image thermal maps of the surface of a concrete member. Voided and delaminated areas are filled with either air or water, each of which has different thermal properties than the concrete, resulting in slight differences in temperature. This method offers the distinct advantage of being performed at a distance from the member examined.

Measuring the electrical properties of reinforced concrete allows an assessment of the rate of corrosion occurring in the reinforcing steel. The two most common tests involve measuring either the potential or the resistivity of the concrete.

- Half-cell – The difference in electrical potential between a copper–copper sulfate half-cell and the steel reinforcing in a reinforced concrete member is measured by placing the half-cell on the surface. Taking measurements across the surface of the member on a grid pattern allows a map of potentials to be created. Typically a measurement that is more negative than -0.35 V is indicative of ongoing corrosion while a measurement less negative than -0.20 V indicates no ongoing corrosion.

- Electrical resistivity – Resistivity is most commonly measured with a Wenner probe that uses four equally-spaced probes. A current is generated in the outer probes and the potential of the resulting field is measured by the inner probes, allowing the resistivity of the underlying concrete to be calculated. The more porous the concrete and the higher the salinity of fluid contained in the pores, the less the resistivity of the concrete. Resistivity values of less than 10 kΩ-cm are indicative of a high rate of corrosion.

Combining maps of existing delaminations and spalls and of active corrosion allow for an assessment of the need for concrete surface repairs and for an estimate of repair quantities for scoped work.

3.5.3 Nondestructive Testing – Steel

Most NDT on in-service steel members concerns finding small cracks before cyclic loading drives the crack size to an unacceptable level. Methods include:

- Eddy current – These are the circular currents induced in the surface of a conductor by the presence of a varying magnetic field. An Eddy current probe uses the field produced by a coil carrying alternating current to produce Eddy currents in a steel member. Cracks in the surface interrupt Eddy currents. By sensing the change in impedance, the Eddy current probe can be used to detect even very small cracks at the surface of a steel member. Although limited to finding surface cracks, this method has the advantages of being quick and requiring little in the way of surface preparation for the examined member.

- Dye penetrant – This is another NDT method limited to detecting surface cracks. The method consists of mechanically cleaning and chemically degreasing the steel surface to be examined, applying a entrant solution that is removed from the surface after a few minutes, then applying a developer that stains any area on which penetrant still remains, that is, cracks in which the penetrant is trapped. This test offers the advantage of being simple to perform and to understand, and delimitates the extent of any surface cracking.
- Magnetic particle – This method uses disturbances in the magnetic field induced in the steel member being examined to attract dyed iron particles to any surface or near-surface cracks. It offers the advantages over the dye penetrant test of requiring less surface preparation and of detecting cracks that are near the surface.
- Ultrasound – Ultrasound testing (UT) is similar to the ultrasonic methods used for concrete. Changes is the ultrasonic waves propagated through the steel are used to detect internal flaws; so while this method does not work well for detecting surface flaws it detects below-surface cracks undetectable by the previous methods. UT can also be used to determine the thickness of a steel member.
- Radiography – The examination of welds by the use of X-rays is common in the fabrication shop but is rarely done on bridges in service due, in part, to the need to handle equipment containing radioisotopes and to shield any adjacent traffic from radiation. However, it may be used to confirm the integrity of critical welding done as part of rehabilitation or repair work.

3.5.4 Sampling

When visual observations and NDT confirm the need to undertake maintenance action on a bridge it may be prudent to take samples to test the properties of the material and the underlying structure of the member.

If a chemical reaction such as an alkali–silica reaction is suspected as the cause of concrete deterioration, the concrete may sampled to conduct chemical analysis. When a concrete bridge deck is chosen for repair work, cores may be taken for physical examination to determine depths of delaminated layers and to examine the integrity of the concrete matrix. If the strength of a concrete member is in doubt, similar cores may be taken to test in a hydraulic press.

Similarly, a coupon may be cut from a low-stress area of a steel member for strength testing. If a proposed repair to a steel member involved welding, the weldability of the existing steel may be confirmed by subjecting small samples to chemical composition tests. Particularly of concern is the carbon content of the existing steel. A recent cautionary tale is the retrofit of the US-77 steel truss bridge over the Canadian River outside of Lexington, Oklahoma in 2013. This resulted in the closing of the bridge after cracks in the steel were discovered. Unknown to the designer of the repair, the steel in the truss was a manganese–alloy steel that was not suitable for welding [13].

While timber bridge members are typically examined solely by visual inspection and physical sounding, if damage from insects or rot is suspected, a small core may be taken to examine the structure of the interior.

3.5.5 Inspecting Fiber-Reinforced Polymer

Fiber-reinforced polymer (FRP) is a material that is becoming more commonly used in bridge construction, both in new construction and in repair work. The three uses of FRP that a bridge inspector is most likely to encounter are:

- Fiber wraps over reinforced or prestressed concrete members to provide either tensile or shear capacity, or to provide confinement for a compression member.
- Fabricated shapes, typically as panels for a deck. Deck panels are typically top and bottom facesheets sandwiched about a core. The core might be a honeycomb of FRP material, foam, or an assembly of pultruded shapes.
- Concrete reinforced with glass fiber-reinforced polymer (GFRP) bars in lieu of the typical steel reinforcing bars.

Fiber wraps are bonded to the surface of existing members by means of a chemical adhesive. Since they are on the surface, wraps are typically accessible for visual inspection. The three components of a wrapped area that may be subject to deterioration and damage are the wrap, the adhesive, and the underlying substrate material of the member. NCHRP Report 564, *Field Inspection of In-Service FRP Decks* lists several visual signs of damage to FRP material that are relevant to wraps. They are:

- Blistering – Due to underlying trapped moisture;
- Discoloration – Possibly due to deterioration from prolonged ultraviolet light exposure or to whitening that occurs due to excessive strains;

- Wrinkling – Occurs during improper layup when installed;
- Exposed fibers – Occurs due to mishandling during installation;
- Cracks or impact holes – These are often tears in the material due to impacts; and
- Scratches – These may be inflicted on the material by mishandling during installation or by external causes afterward.

More common than failure of the wrap material is failure of the adhesive. Proper adhesion requires proper surface preparation, the results of the failure to do so might not be immediately apparent after installation. Adhesive failure may manifest itself as loss of bond at the edges of the attachment of the wrap.

Although failure of the substrate material may not be visible through the overlying FRP wrap, the integrity of the underlying concrete may be checked by physically sounding with a hammer just as with an exposed concrete surface. A solid surface will have a higher pitched ring than an area of delamination that will produce a rattling sound similar to exposed concrete.

The surface of deck panels may be inspected in a manner similar to that used to inspect fiber wrap, with the addition of paying particular attention to review the effects of abrasion from traffic on the wearing surface. With FRP deck panels, there are several mechanical connections to examine for damage or loosening. These include: key joints between panels, the attachment of the panels to the girders, and sometimes of the rail to the exterior deck panel.

Concrete members with GFRP reinforcing can be inspected by the same methods that either an unreinforced or a conventionally reinforced concrete member (e.g., with steel) would be. The exception is that since GFRP bars are nonmagnetic, they cannot be detected with a pachometer.

Given the relative newness of FRP material in bridge construction, and their use as a material for repairs and retrofit, it would behoove those involved in scoping and scheduling bridge maintenance work to pay particular attention to the performance on structures utilizing FRP in their bridge inventory.

3.5.6 Posttensioning Ducts

Posttensioning has allowed for the use of concrete as the primary superstructure material for bridges as span lengths have increased over time. However, the durability of posttension tendons became a concern in the United State after the discovery of significant corrosion and the failure of tendons on two major bridges in Florida, first in 1999, then in 2000. These discoveries came after some notable bridge failures in Europe in the 1980s and 1990s due to corrosion of posttension tendons.

Many of these failures were due to the presence of voids in what should have been completely grouted posttensioning ducts and anchorages. These issues were addressed for subsequent construction with improved grouting techniques (such as vacuum-assisted grouting), the use of prepackaged grouts and the use of plastic ducts (which are more air tight than metal ducts). However, in the late 2000s tendon failures were again found, this time from higher-than-expected levels of chlorides and sulfates in some of the prepackaged grout. In 2011, the FHWA issued a memorandum advising the states about a particular prepackaged grout of concern for excessive chlorides. As a result of these issues, in October 2013, the FHWA published *Guidelines for Sampling, Assessing, and Restoring Defective Grout in Prestressed Concrete Bridge Post-Tensioning Ducts.*

Posttensioned bridges in service should have a special inspection to check ducts and anchorages for voids, which is more of a concern for bridges constructed prior to the early 2000s, and for issues with the grout. A typical investigation may consist of the following steps:

1. Locate existing posttensioned ducts by NDT methods, possibly using GPR.
2. Using UT, examine posttensioned ducts for voids.
3. Confirm the results of UT by drilling into some of the voided ducts and investigating visually with a borescope.

Following the procedures in the FHWA guide, sample the grout in the posttensioned ducts at selected locations. The guide recommends the number of samples to be taken and the location of the samples

3.5.7 Structural Health Monitoring

Structural health monitoring (SHM) refers to the use of sensors to remotely report the response of a bridge structure to external loadings and environmental conditions. SHM makes use of a number of the same technologies used to test materials and structures in the laboratory including sensors for strain, displacement, temperature, inclination, acceleration, and pressure sensors. Dong et al prepared a synthesis of SHM technologies for used by the Alaska DOT including a discussion of available sensors and proprietary systems available for use on bridges [14].

Often SHM technologies are applied during the construction of bridges to validate design assumptions and to ensure compliance with contract specifications for the materials. Examples include load cell tests to confirm the capacity of drilled shaft foundations and temperature monitoring of the curing of large-volume concrete elements to ensure mass concrete placement specifications are followed. For bridge maintenance applications, SHM has been typically been applied to pursue one of two goals: either to calibrate design assumptions to provide for a more accurate analysis of structural capacity, or to alert bridge owners of a detrimental change in bridge condition for a particular failure mode of concern.

In 2009 the Engineer Research and Development Center, Construction Engineering Research Laboratory of the US Army of Corps of Engineers implemented a project to test the use of a suite of SHM technologies on long steel truss bridges [15]. Two bridges over the Mississippi River were used as test subjects, the Government Bridge (also known as the Arsenal Bridge) connecting Rock Island, Illinois and Davenport, Iowa and the I-20 Bridge connecting Vicksburg, Mississippi, and the State of Louisiana. The sensors installed on the Government Bridge included: strain gages, accelerometers, corrosion meters, and (to detect crack growth) acoustic emission sensors. The sensors installed the following year on the I-20 Bridge included: strain gages, accelerometers, corrosion meters, tilt sensors, and displacement gages. The information from the accelerometers and strain gages were used in finite element models of the superstructures of the bridge providing the owners with improved estimates of the structural capacities of the bridges. This proved particularly useful when, in April 2011, a 30-ton grain barge struck the I-20 Bridge, forcing a closure of the bridge for several hours while its structural integrity could be confirmed. With the SHM system in place, this work could begin before Louisiana Department of Transportation inspectors arrived on site [16].

A review of SHM projects intended to provide continuous, long-term monitoring of bridge condition and structural performance shows similar results to those reported by the Corps – that such systems provide useful information but are expensive to install and to maintain, and require specialized personnel [17]. Problems include maintaining power supplies, maintaining the communication links, and the difficulty in replacing sensors.

There are other methods of monitoring the conditions at bridge sites that do not directly measure structural response, yet provide information regarding the need for bridge maintenance actions. Two examples in common use include traffic cameras and stream gages. The cameras used to provide information to traffic operations centers common in major metropolitan areas can also, with sufficient resolution, capture information concerning the two largest issues for preventative maintenance in bridges – drainage and traffic. Clogged deck drains and scuppers will result in slow drainage and, possibly, ponding of water on bridge decks during precipitation events. Traffic collisions with bridges will be captured on video if directly in the field of view of a camera, or may be indirectly inferred shortly after an incident from disruption to the usual traffic patterns. Receiving notice of bridge impacts as soon as possible helps first responders arrive as quickly as possible to address fires or chemical spills that may damage bridge structures, and also facilitates the soonest possible inspection of a bridge to determine if its integrity has been compromised to reduce the risks to the traveling public.

As discussed in Ref. [18], the scouring of bridge foundations is one of the largest causes of bridge failures. The United States Geological Survey (USGS) maintains a system of approximately 8000 active stream gages in the United States. Real-time information from these gages is available from the USGS web page: http://waterdata.usgs.gov/nwis/rt. Many bridge owners monitor these gages during large precipitation events and use the information in their scour action plans required for bridges designated as scour critical. A sufficiently large rainfall in the watershed upstream of a scour critical bridge will trigger further investigation at the site to determine if the bridge is safe to remain open.

3.5.8 Sonar and Underwater Inspection

As per the NBIS, the submerged portions of bridge substructure elements must be inspected at regular intervals not to exceed 60 months. In clear water, this requires a visual inspection; but, as is far more typically the case, it requires an inspector to feel the substructure element above the mudline to insure the continuity and integrity of the element and to look for any exposure of foundation elements, such as pilings, that would be indicative of the occurrence of scouring [19].

Typically, while the inspection of the underwater bridge elements is conducted, the bottom of the channel at the bridge site is mapped to provide for a scour analysis of the location. For shallow streams, this can be conducted by an inspector in waders with a survey rod. For locations under a few feet of water, diving equipment is required. To minimize the amount of diving required (with its adjacent costs and risks), sonar depth finding equipment (fathometer) has been used for many years to map channel bottom elevations during bridge inspections. However, since a fathometer is limited to determining the depth immediately below the device, it cannot be used to image the submerged substructure elements.

To get an image of the bottom of a channel that includes the submerged bridge elements, either side-scan sonar, or multibeam echo procedures must be used. A side-scan sonar device sends sound energy out in a fan shape rather than a directed beam toward the bottom. An object on the bottom of the channel creates both a strong return signal and an area of shadow with little return signal strength. This allows for a detailed image of the bottom; however, other techniques must be used to establish the depth to the bottom.

A multibeam echo sounder sends out multiple beams at various angles. The result is more resolution of the bottom, but the greater amount of signal results in less-apparent contrast. The multibeam image requires significant postprocessing to produce an image which makes it less useful for real-time imaging. A side-scan sonar unit would be preferable when an immediate evaluation of the condition of a bridge pier during a high flow event. However, the greater resolution of a multibeam image is of use for scheduled bathymetric surveys. During the period 2010–2014, the USGS conducted bathymetric surveys of bridges over the Missouri River for the Missouri Department of Transportation. During construction of the new Amelia Earhart Bridge over the Missouri River at Atchison, Kansas, the USGS conducted a series of bathymetric surveys to monitor the condition of the existing bridge in place during construction of the new. During construction, the second and the third (each in a separate year) highest stream flows recorded by the stream gage at Atchison were logged. The results from these investigations allow for improved hydraulic modeling of the bridge sites while eliminating the need for divers.

3.6 LOAD RATING

A load rating is a determination of the safe-load-carrying capacity of a highway bridge in terms of a particular vehicular loading. The vehicular loading is usually an assumed truck, defined by axle load and spacing, which may or may not be combined with uniform load loading to represent lighter passenger automobile traffic. Though it is a simple concept, the analysis involved can be quite significant. In the United States, the *AASHTO Manual for Bridge Evaluation* is the guiding specification for highway bridge load ratings and contains a number of illustrative examples. The details of the actual analysis steps are not discussed here, but the general concepts are reviewed so that applicability of the results of load rating analysis for scheduling and scoping bridge maintenance actions can be discussed.

3.6.1 General Approach

Load ratings results are reported in terms of ratings and rating factors (RFs). An RF is the ratio of the available live load capacity of a bridge to the live load demand imposed by the truck (live) loading assumed in the analysis. The live load capacity of a bridge at any point in time is the total capacity of the bridge as originally constructed, less any capacity lost to deterioration or damage to the structure and less the capacity used to support the dead load of the structure. An RF less than 1.0 indicates that there is less capacity than demand. Impact magnification is generally applied to the truck loading and safety factors are applied to the dead and to the live loadings. The general form of the equation is:

$$RF = \frac{C - A_1(DL)}{A_2(LL)(1 + I)}$$

Equation 3.2 General Form of the Load Rating Equation.

Where, C, current capacity of the structure; DL, dead (or other permanent) load demand; LL, live load demand; I, impact factor; A_1, magnification factors applied to dead load effects; and A_2, magnification factors applied to live load effects.

During a load rating analysis, an RF can be calculated for each load effect (shear, moment, etc.) on each element for each truck loading under consideration. The smallest RF is the controlling RF for truck loading for that bridge. The value given for the rating for each truck loading considered is the RF multiplied by the gross weight (in tons) of the vehicle used in the analysis. For example, if the lowest RF for a steel girder bridge was 0.95 due to moment on an interior girder line for a HL-93 loading, the rating for the bridge for that load would be 0.95 × 36 tons = 34.2 tons.

Calculating the RF for each load effect on each element can be laborious. Experienced bridge designers typically know which load effects on which elements will be critical for a given bridge types and the range of typical truck loadings. Guidance is given the Manual for Bridge Evaluation on what bridge elements will typically not control a load rating analysis.

3.6.2 Analysis Methodologies

The purpose of a structural analysis is to determine whether or not a structure has sufficient capacity to bear the demands imposed by an assumed loading. How well the mathematical model used in such an analysis reflects the behavior of the real world structure is constrained by the accuracy of the method and by the uncertainty regarding the assumed parameter values used in the model. Factors are applied to parameters on the demand and/ or the capacity sides of the analysis to provide confidence that the results of an analysis reflect reliably safe limits for the loading of the structure. To conservatively estimate a safe operating load, the assumed capacity may be reduced and/or the assumed demand may be increased beyond the values actually expected.

For highway bridge design in the United States, there are three design/ analysis methodologies in place reflecting the evolution of design philosophies in structural engineering over the course of the past century. The earliest design methodology used was allowable stress design (ASD), alternately known as working stress design. In this method, the expected load demand and capacity for a structural member are calculated. A single factor of safety (FS) is used to reduce the assumed capacity of the structure in the analysis.

$$\text{Demand} \leq \frac{\text{Capacity}}{\text{FS}}$$

Equation 3.3 General Form of the Allowable Stress Equation.

There are different FS for different materials and load effects, reflecting to some degree how well the properties are known for a particular material and the severity of the consequence of failure for the load effect. A shear failure in a concrete member would be more sudden and complete than a bending failure in a steel member; hence, a larger FS would be used.

The next methodology, which was first published in the 1971 Interim Specification as an alternate specification, is load factor design (LFD). LFD recognizes that there are different degrees of certainty regarding different sources of loading, that is, the dead load from the self-weight of a structure is known to a greater degree of certainty than are those for either live or environmental loads. Accordingly the load factors for dead loads are less than those for either live or environmental loads. Reduction factors are also applied to member capacities, although the reduction factors applied to some common member capacities (such as steel in bending) are 1.0.

$$\Phi \, \text{Capacity} \geq \gamma [\beta_{DL} \, DL + \beta_{LL} (LL + I)]$$

Equation 3.4 General Form of the Load Factor Equation.

Where, Φ, reduction factor applied to capacity; γ, load factor for loading combination; β_{DL}, dead load factor; β_{LL}, live load factor; DL, dead load effect; LL, live load effect; and I, impact.

On the capacity side, LFD takes advantage of the fact that a structural member does not deform when the stresses at the extreme fiber reaches yield, but only after the entire cross-section yields, that is, once the section goes plastic. When calculating capacity for most load effects, the plastic capacity of members is utilized. The majority of the bridge inventory in the United States constructed since the early 1970s was designed utilizing LFD.

The methodology currently used in bridge design is load and resistance factor design (LRFD). LRFD was first adopted as an alternative design specification by AASHTO in 1993. FHWA required that all bridge designs initiated after October 2007 (and all culvert designs after October 2010) utilize LRFD for projects receiving federal aid. Most current bridge design methodologies in Europe, Canada, and Australia are LRFD methodologies. Similar to LFD, LRFD utilizes separate modification factors for loads and for resistance (capacity) and utilizes the plastic capacity of members for most load effects. LRFD factors, though, have been calibrated to a target level of reliability based on statistical studies of the variability of loadings and of material strengths. The result is to provide a more uniform margin of safety for differing structure types and materials, and to provide a rational method of establishing factors as new materials and structure types are incorporated into highway bridge construction.

$$\Phi R_n \geq \Sigma \eta_i \gamma_i Q_i$$

Equation 3.5 General Form of the Load and Resistance Factor Equation.

Where, Φ, reduction factor applied to resistance (capacity); R, nominal resistance; η_i, load modified relating to ductility, redundancy or operational classification; γ_i, load factor for loading i; and Q_i, load effect i.

The FHWA requires all bridges designed with LRFD after 2010 to be load rated using load and resistance factor rating (LRFR) specifications. Bridges which were designed with ASD, and are not deficient (either SD or FO), may be load rated by allowable stress rating (ASR). Effectively, most bridges then may be load rated utilizing the methodology used in the original design or any methodology subsequently adopted. However, since the requirement to report a load rating to the FHWA started with the inception of the NBIS in 1971, at approximately the same time the LFD specifications were adapted, most existing bridges have been rated utilizing load factor rating (LFR).

In any of the three analysis methodologies, the first load rating analysis for a bridge is to review it for the design truck loading at the inventory and operating load levels. The inventory rating is intended to reflect the weight of truck that can be driven over a bridge indefinitely without damage to the structure. The operating rating reflects the largest weight of truck that can be safely driven over a bridge on occasion. Routine truck traffic at the operating level would be expected to shorten the life of the bridge. For ASR, the inventory loading is that which results in a maximum stress in the structure equal to 55% of the yield stress; the operating loading results in 75% of the yield stress. In calculations using LFR, the live load multiplication factor is 2.17 for inventory ratings and 1.30 for operating ratings. For LRFR, utilizing the strength I loading combination, the live load factor applied to inventory calculations is 1.75 and for operating calculations it is 1.35.

Design truck loads (HS20-44 for ASR and LFR, and HL-93 for LRFR) are intended not to reflect any particular real-world truck and vehicular loading, but to produce an envelope of load effects for the analysis of the structure that would encompass the myriad of likely truck and vehicle combinations on the real bridge (Figure 3.5). The design loadings incorporate other uniform and axle loads, either in combination with or alternatively to, the assumed semi-trailer truck loading. A bridge that has an inventory RF ≥ 1.0 for either the HS20-44 or the HL-93 design loading can be considered adequate for all legal truck loadings. For a bridge to have an inventory RF < 1.0, either the loading used in the original design was less than the current design loading, or the condition of the bridge has deteriorated and diminished its load capacity.

A bridge with an inventory RF < 1.0 and an operating RF ≥ 1.0 for the design truck should be able to carry the legal trucks defined by AASHTO, but may not be able to carry some of the heavier state legal trucks and should be load rated for both the AASHTO and state legal trucks.

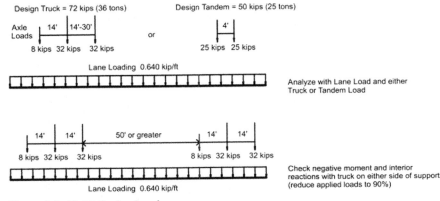

Figure 3.5 *HL-93 Design Load.*

3.6.3 Truck Loadings

With some exceptions, maximum gross vehicular weight (GVW) for un-restricted operation on the United States Interstate Highway System is 80,000 lbs. Off of the Interstate system, the individual states are free to set maximum vehicular weights for unrestricted operation. Fourteen of the states allow for heavier trucks with six of those states allowing for over 100,000 lbs [20]. For comparison, the semi-trailer truck assumed in the HS20-44 and HL–93 design loads has a GVW of 72,000 lbs.

To account for these larger trucks, AASHTO and the states define legal trucks. These loadings consist only of the truck axle loadings, without any accompanying or alternate uniform loadings in the analysis. AASHTO de-fines the three type 3 trucks shown in Figure 3.6. States may develop their own legal trucks for analysis, indicative of their own traffic characteristics and laws.

If a bridge has an operating RF > 1.0 for one or more of the legal trucks, it should be posted to restrict the weight of truck traffic on the

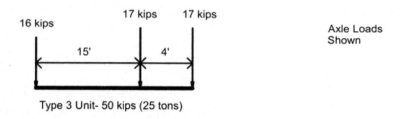

Type 3 Unit- 50 kips (25 tons)

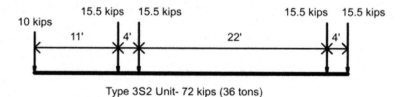

Type 3S2 Unit- 72 kips (36 tons)

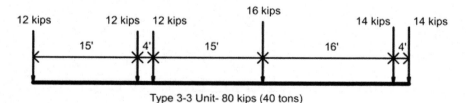

Type 3-3 Unit- 80 kips (40 tons)

Figure 3.6 *AASHTO Type 3 Legal Load Trucks.*

bridge. The posted weight should not exceed the operating rating for the bridge, but a bridge owner may choose a more restrictive posting out of concerns for further deterioration of the structure.

The type 3 truck loadings were developed to produce worst-case demands by legal trucks. The smallest truck usually produced the most demand on shorter span bridges while the medium length and the longest trucks governed for medium and longer spans, respectively. However, the trucking industry has developed other legal axle configurations, particularly for specialized hauling vehicles (SHV), which produce more severe loading on some bridge span arrangements. The single-unit, short wheelbase, multiple-axle trucks are used in construction, waste management, and bulk commodity hauling. NCHRP Project 12-63 studied contemporary truck configurations and found the need to develop a series of single unit SHV legal trucks, SU4 to SU7 (Figure 3.7).

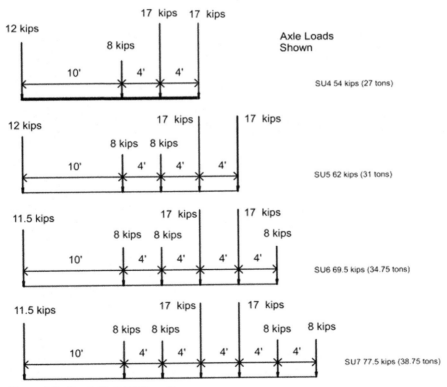

Figure 3.7 *Single Unit SHV Loadings.*

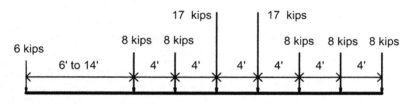

Axle Gage = 6'0"
Gross Vehicle Weight = 80 kips (40 tons)

Neglect any axles that do not contribute to
the maximum load effect.

Figure 3.8 *Notional Rating Load.*

As a screening tool, another loading, the notional rating load (NRL) was proposed and is included the current Manual for Bridge Evaluation (Figure 3.8). This loading does not represent any particular truck, but is meant to result in load effects that envelop those of most SHV. A bridge with a RF ≥ 1.0 for the NRL would not need to be checked for the SU4–SU7 trucks.

3.6.4 Load Rating by Testing

Accuracy in a load-rating analysis is contingent on having accurate and complete information about the structures as it was built and about its current condition. Unfortunately, complete design plans are not available for all bridges, particularly older ones. Depending on the amount of information available, it may not be possible to develop an analytical model for structural analysis or several assumptions may have to be made in development of the model. These bridges may be rated for their safe-load-carrying capacity by testing the structures with load tests. Such tests consist of applying known loads to the structure and measuring its response with NDT methods.

The procedures for conducting such tests and for determining a rating from the results were originally described in the *Manual for Bridge Rating through Load Testing*, published in NCHRP Research Results Digest #224 [21]; and updated for LRFR in Section 8 "Nondestructive Load Test" of the *Manual for Bridge Evaluation*. Load tests are classified into two types, diagnostic and proof. A proof test involves applying incrementally increasing loads to a structure until a predetermined safe load has been reached or the bridge begins to exhibit inelastic behavior. A diagnostic test uses a lower magnitude of load than does the proof test. The responses measured from a diagnostic test are used to validate and calibrate a structural analysis model of the structure.

Even when an analytic model is based on relatively complete plans, the model may be overly conservative in its predictions about the behavior of the real bridge. Factors listed in the Manual for Bridge Evaluation, which may provide additional capacity unaccounted for in a model include:

- Unintended composite action between the girders and the deck;
- Unintended continuity and fixity provided by the slab or by frozen bearings;
- The participation of secondary members in increasing the stiffness of the structure;
- The participation of nonstructural members such as parapets and barriers; and
- The portion of the load carried by the deck.

By using load testing to determine a more accurate rating for a bridge in good condition, but a light rating, maintenance resources can be better allocated to actual needs.

3.6.5 Fatigue Evaluation of Steel Bridges

Bridges with steel superstructures are subject to the effects of fatigue from cyclic tensile loads. Damage from fatigue may be broadly divided into two categories, load-induced and distortion-induced. Load-induced damage is that resulting from in-plane stresses in the plates that comprise the cross-section of a steel member. Distortion-induced damage is due to secondary stresses, typically at the attachment of secondary framing members (diaphragms and cross-frames) to a primary structural member. Chapter 7 of the AASHTO *Manual for Bridge Evaluation* provides for the evaluation of load-induced fatigue damage. The manual considers evaluation of distortion-induced damage as beyond the scope of typical bridge evolution due to the need for advanced structural analysis techniques.

The evaluation of the potential for load-induced damage is typically more of a concern for bridges designed prior to the issuing of the 1974 Interims of the AASHTO *Standard Specifications for Bridge Design*, than it is for contemporarily designed bridges. The interims placed typical steel bridge details into categories with corresponding *S–N* (stress ranges vs. cycles to failure) curves. This resulted in the virtual elimination of some problem details, such as cover plates terminating in welds in tension regions. In 1974, requirements for minimum Charpy test values were also implemented for bridges. The Charpy test is a measurement of the toughness of material and, by extension, its resistance to crack propagation.

Distortion-induced damage is a concern for bridges designed prior to 1985. At that time the specifications were updated to require a positive attachment between connection plate stiffeners and girder flanges. This is discussed further in Chapter 5.

An evaluation of existing bridge details for load–induced fatigue damage is made only if the detail is not cracked. If the detail is cracked, it is presumed that its fatigue life has been exhausted and should be repaired. An uncracked fatigue prone detail is evaluated first to see if it will meet the criteria for infinite life. A detail may be assumed to operate throughout the expected life of a bridge if the twice the effective stress range experienced by the detail is less than the constant-amplitude fatigue limit given for that detail category in the AASHTO LRFD Bridge Design Specifications. The effective stress range at the detail is calculated as:

$$\text{eff}(\Delta f_s) = R_s \Delta f$$

Where R_s is the stress–range estimate partial load factor, a value of 1.0–0.85, dependent on the source of the stress range estimate; and the stress range, Δf, is determined by either direct measurement with strain gages, or is calculated using the LRFD fatigue truck or a fatigue truck determined by weight-in-motion studies at the site.

Should the detail not satisfy the infinite life criteria, the finite life expected for the detail is calculated from the equation:

$$Y = \frac{R_R A}{365 n (\text{AADT})_{SL} [(\Delta f)_{eff}]^3}$$

Where, Y, fatigue life in years; R_R = resistance factor for evaluation; A, detail category constant from LRFD Bridge Design Specification; n, number of stress range cycles per truck passage; AADT_{SL}, average number of trucks per day in a single lane averaged over the fatigue life of the bridge; Δf_{eff}, effective stress range.

3.6.6 Programming Maintenance Actions

The goal of load rating is to determine the safe-load-carrying capacity of highway bridges. If that capacity is less than the demand imposed by legal truck traffic, then access to the bridge is restricted to lighter traffic by posting. Restricting the weight of trucks allowed on a bridge has economic consequences due to increase in costs, from fuel usage and longer travel times, to commercial activities that depend on trucking. If a posted bridge is the only highway access into a community, the impact of restricting the flow of goods can be significant. Posted bridges are a high priority when scheduling bridge maintenance projects.

The results of a load rating can help an engineer in scoping repair work. If a deficient rating is due to a truck with a shorter wheelbase, such as the type 3 unit as opposed to the type 3-3 unit truck, the deficiency may be in the shear capacity of the superstructure members rather than the flexural capacity (Table 3.14).

REFERENCES

[1] O'Connor DS, Hyman WA. Bridge management systems. Washington, DC: Federal Highway Administration, US Department of Transportation; 1989. FHWA-DP-71-01R.

[2] Thompson PD, Shepard, RW, Scotsdale, AZ. AASHTO commonly-recognized bridge elements, successful applications and lessons learned. Summary of the proceedings of the national workshop on commonly recognized measures for maintenance; 2000.

[3] Khanzada MK. State of bridge management in Canada. Fargo, ND: North Dakota State University of Agriculture and Applied Science; 2012.

[4] Bisby, LA, Briglio, MB. ISIS Canada Educational Module No. 5: an introduction to structural health monitoring. s.l.: ISIS Canada; 2004.

[5] Hearn G, et al. Bridge preservation and maintenance in Europe and South Africa. Washington, DC: Federal Highway Administration; 2005. FHWA-PL-05-002.

[6] Hearn G. NCHRP synthesis 375 bridge inspection practices. Washington, DC: Transportation Research Board; 2007.

[7] Moss, J. Bridge Inspection Competence and Training – Phase II. Birmingham: Atkins; 2012. Document No. 5094334/017.

[8] Everett TD, Weykamp P, Capers HA. Bridge evaluation quality assurance in Europe. Washington, DC: Federal Highway Administration, US Department of Transportation; 2008. FHWA-PL-08-016.

[9] Washer G, Connor R, et al. NCHRP report 782 proposed guideline for reliability-based bridge inspection practices. Washington, DC: Transportation Research Board; 2012.

[10] Reising RS, Connor RJ, Lloyd JB. Risk-based bridge inspection practice. Washington, DC: FHWA; 2014. FHWA/IN/JTRP-2014/11.

[11] Carino, N.J. The impact-echo method: an overview. Proceedings of the 2001 structures congress & exposition. Washington, DC; 2001.

[12] Gucunski N, et al. Nondestructive testing to identify concrete bridge deck deterioration. Washington, DC: Transportation Research Board; 2013. SHRP 2 Report S2-R06A-RR-1.

[13] Two cities now one community again as Purcell/Lexington bridge reopens. ODOT News. [Online] 2013. Available from: http://www.okladot.state.ok.us/newsmedia/press/2014/14-026_Two_cities_now_one_community_again_as_Purcell_Lexington_bridge_reopens.pdf. PR# 14-026

[14] Dong Y, Song R, Liu H. Bridges structural health monitoring and deterioration detection synthesis of knowledge and technology. Fairbanks, AK: Alaska University Transportation Center; 2010. INE/AUTC #10.06.

[15] Implementation of a novel structural health management system for steel bridges. U.S. Army Corps of Engineers (ERDC-CERL). Huntsville, AL 2010 U.S. Army Corrosion Summit; 2010.

[16] Holland, M. Bridge-monitoring technology plays critical role following barge crash. CorrDefense. [Online] 2011. Available from: http://corrdefense.nace.org/corrdefense_summer_2011/project_news2.asp

[17] Sweeney, S. Observations from structural health monitoring of a steel truss bridge. Standing Committee on Bridges and Structures presentation. [Online] 2010. Available from: http://bridges.transportation.org/Documents/2010_SCOBS_presentations/US_Army_Corp_-_Structural_Monitoring_Systems.pdf

[18] Collins Engineering. Underwater bridge inspection. Washington, DC: Federal Highway Administration; 2010. FHWA-NHI-10-027.

[19] Kansas Department of Transportation Bridge Office. Bridge inspection manual. Topeka, KS: Kansas Department of Transportation; 2013.

[20] US Department of Transportation. Comprehensive truck size and weight study. Washington, DC: Federal Highway Administration; 2000. FHWA-PL-00-029.

[21] NCHRP. Research results digest – manual for bridge rating through load tests. Washington, DC: Transportation Research Board; 1998. 224.

CHAPTER 4

Preventative Maintenance

Overview

The role of preventative maintenance in the cost-effective management of bridges is discussed in this chapter. Challenges to quantify the cost benefit from preventative maintenance are examined. The role of maintenance inspections is reviewed and particular preventative maintenance activities for specific bridge elements are examined.

4.1 INTRODUCTION

In Chapter 1 of this text, bridge maintenance actions were distinguished as either preventative or substantial. Preventative maintenance actions are those undertaken to prevent or mitigate deterioration, while substantial maintenance actions are those undertaken to repair damage. Substantial maintenance is conducted to improve the condition of bridge components, while, ideally, preventative maintenance is conducted on components in good condition with the goal of maintaining that condition. Substantial maintenance is reactive in nature, and preservation is proactive. Viewing it in terms of the condition of the bridge over its life, preventative maintenance is a series of small activities to maintain a consistent condition.

The American Association of State Highway and Transportation Officials (AASHTO) Subcommittee on Maintenance defines preventative maintenance as "a planned strategy of cost-effective treatments to an existing roadway system and its appurtenances that preserves the system, retards future deterioration, and maintains or improves the functional condition of the system (without substantially increasing structural capacity)". Preventative maintenance work is ongoing, planned, and undertaken with the intent to maintain a bridge as fully operational for the lowest cost over its life cycle. The activities are usually small scale and are often done by the owner's in-house staff.

The *AASHTO Maintenance Manual for Roadways and Bridges* further divides preventative maintenance activities into two categories: scheduled and response [1]. Scheduled activities are conducted on a regular basis with consistent intervals in between. Typically scheduled activities include cleaning and sealing concrete decks and cleaning out bridge drains and scuppers. Activities done in response are those identified by inspections (either formal bridge inspections or maintenance inspections). Although many

Highway Bridge Maintenance Planning and Scheduling
http://dx.doi.org/10.1016/B978-0-12-802069-2.00004-0

maintenance actions that are required in response to conditions found by inspection will be substantial maintenance, there are several actions such as removing debris or replacing damaged signs that fall under preventative maintenance.

Preventative maintenance activities on bridges are typically directed to mitigating either of the two greatest threats to bridge health: water and traffic. Water can facilitate corrosion and the deterioration of material, can carry away sediment at the bridge approach and in the channel below, can carry debris and can facilitate the growth of biological agents. Errant traffic presents a danger due to collisions with the structure. Excessively heavy trucks can overload bridge elements and the impact forces of vehicles traversing a potholed wearing surface may accelerate wear to the deck and to expansion joints. Preventative maintenance actions commonly are directed to one of the following goals:

1. Ensuring proper drainage of water from the roadway down to the ground line below and away from the bridge site;
2. Sealing the materials of structural elements from exposure to water;
3. Maintaining an adequate and clear waterway under the bridge at water crossings; and
4. Providing proper guidance for drivers on the roadway over the bridge, and at grade separations, underneath the bridge.

4.2 COST EFFECTIVENESS

Preventative maintenance activities are meant to be cost effective, not only in terms of lower cost in terms of materials and labor expending on keeping a bridge in operation over its life, but also in terms of minimizing costs to the users of the facility. A bridge replacement requires significant disruption to traffic as the existing bridge is removed and the new one constructed. Substantial maintenance work to repair or rehabilitate a structure also has significant impact due to lane closures needed to allow for the workers and heavy equipment to remove and replace deteriorated portions of a bridge. In contrast, preventative maintenance actions are quick, typically requiring no more than temporary lane drops to allow crews to conduct the activity and immediately move on. Assuming that the preventative maintenance actions are successful in prolonging the life of the structure and in minimizing the number of substantial maintenance actions that will be required to restore condition, the delay to users is minimized, resulting in reduced travel times and fuel costs.

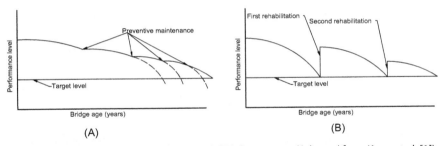

Figure 4.1 *Preventative Versus Substantial Maintenance. (Adapted from Kong et al. [1]).*

Several researchers have proposed numerical models of bridge deterioration and methods of life cycle costings. These will be discussed in Chapter 6 of this text. Such models, though, are limited in their ability to determine the actual cost effectiveness of preventative maintenance, as defined in this text, on the lifetime cost of bridge operations. These models include substantial maintenance activities in their analyses of the effectiveness of maintenance operations. As illustrated in Figure 4.1B, substantial maintenance work restores the condition of a bridge element to its original, or some acceptable state. If no data is available to predict the rate of deterioration of the element after the repair, then the previous rate of deterioration may be assumed for the analysis.

However, preventative maintenance actions work to reduce the rate of deterioration rather than substantially improving the condition of bridge elements, as illustrated in Figure 4.1A. By preventing water and its accompanying debris and salts from infiltrating into the material of the deck and superstructure, or into the mechanisms of the expansion joints and bearing, the rate of deterioration is minimized. So, too, are the impacts of traffic loadings minimized by maintaining a smooth wearing surface on the bridge and on the approaching roadway.

In a 1998 study for the New York City Department of Transportation, Yanev identified the average bridge deterioration rate as the greatest factor in his model of an average bridge condition rating for the system [2]. He assumed that the average bridge deterioration rate was directly influenced by bridge maintenance practices, including preventative maintenance. In the years since, though, there has been little empirical research to precisely quantify the effect of preventative maintenance actions alone on deterioration rates. Even without precise quantification of cost and benefit, preventative maintenance actions still represent good practice for bridge owners in that they facilitate operation of structures at the best level of service. Proper drainage and smooth wearing surfaces encourage steady traffic flows. Good delineation of lanes and clear guidance to drivers encourages the safe flow of traffic.

4.3 MAINTENANCE INSPECTIONS

Bridge inspection is key to the bridge preservation process. It is not a function that should be limited to the scheduled observations and examined by certified inspection personnel every few years or months. Preventative maintenance work is typically performed by in-house staff assigned to maintaining operations on a particular section of highway. These personnel are generally those in most frequent contact with any particular highway bridge and are those best situated to observe, not only the first signs of deterioration, but the conditions that might facilitate that deterioration.

Many highway agencies required maintenance supervisors to visually examine all structures in their assigned territory at regular intervals. As one example, the Texas DOT requires this of their bridge maintenance supervisors every 6 months [3]. A number of guides and reference materials exist, including the Federal Highway Administration (FHWA)'s *Bridge Maintenance Training Reference Manual*, which may be used to guide and instruct personnel on specific maintenance inspection procedures.

Regardless of the specific procedure, any inspection of a bridge structure will involve four sets of observations for each element of the bridge. Those are that of the:

1. Condition of the material of the element;
2. Integrity of any protective system;
3. Integrity of the structure of the element; and
4. Orientation/alignment of the element.

As discussed in Chapter 2, each material used in bridge construction is subject to deterioration. Concrete may crack from chemical deterioration or from corrosion of reinforcing steel. Steel may corrode and timber may split or rot. Any inspection should involve checking visible surfaces for signs of deterioration.

Many elements have protective systems, the most obvious of which is the painting of structural steel, in place to forestall deterioration. Other protective systems include deck overlays and the armoring of berms with either rock or concrete. A primary focus of preventative maintenance is to maintain the effectiveness of these systems.

Any concerns about structural adequacy should trigger an inspection by certified bridge inspection staff. However, a maintenance inspection is not only concerned with the integrity of major structural elements, but also of elements such as expansion joints. Maintenance staff should be on the lookout for cracks and loss of section in load-carrying members such as girders and tears in expansion joints. Similarly, elements should be oriented and aligned as expected. The vertical surfaces of substructure elements such as pier and abutment backwalls should be plumb. Bearings should be in either the expanded or contracted position correct relative to the ambient temperature at the time of observation. Expansion joints and approaches should be flush with the wearing surface of the deck.

In addition to observations particular to each element on the bridge, the maintenance inspector should look at the bridge site as a whole and ascertain that at the site there is:

1. Positive drainage away from the bridge and the approaches;
2. Positive guidance for drivers on the roadway on the bridge and, if applicable, under the bridge; and
3. No impediment to the free and smooth flow of water in any channel under the bridge.

Ponded water on a bridge deck may indicate plugged drains. Erosion around the wings of an abutment may indicate a lack of grading away from the approaches. In either case simple maintenance actions may be employed to restore proper drainage and minimize the deterious effects of water at the structure.

Positive guidance for drivers includes clear delination of the roadway and of the structure. It also includes signing to alert drivers to conditions that might violate his or her expectations, such as low overhead clearances. There are few countermeasures that exist to prevent collisions by drivers with bridges outside of guard fence and sand barrels. However, even collisions with those devices incur significant cost for the driver and the bridge owner. Helping the driver in the driving task is the most cost-effective way to minimize collision costs.

As discussed in Chapter 3, a major cause of bridge failures is scour. After high flows recede, there may be evidence of scouring in the form of channel degradation around substructure elements. Also, the accumlation of debris from high flow events may excerbate scour effects by forcing the flow of water through a partial section of the intended waterway.

The sections below discuss particular deficiencies that may be encountered at locations on a bridge and particular preventative maintenance actions to address them and to slow the rate of deterioration of bridge elements from other causes.

4.4 BRIDGE DECKS AND EXPANSION JOINTS

4.4.1 Deck Drainage

As the component which both directly bears traffic loads and conveys run-off from precipitation, bridge decks require the most attention from preventative maintenance efforts. Intact bridge decks and expansion joints also serve to shield the structural elements beneath from runoff which may be laden with salts and debris (Figure 4.2). The first priority in maintaining the deck is to provide positive drainage by maintaining open drains and scuppers. Drains can become clogged with sand and other debris from the roadside. Maintenance crews can clean out smaller debris by pushing it through the system with water under high pressure. Larger items, such as cans and bottles, may have to be snaked out from above, or from turnouts in the drainpipe if so provided (Figure 4.3).

Bridge decks which have drainage over the side through openings in the bridge rail and shorter decks which convey all of their drainage to the end of the bridge still require attention in maintaining gutter lines free from obstructions. Once past the end of the deck, a positive path should be provided from drainage down the gutterline of the approach and down the roadside berm. Off of the bridge this may require clearing vegetation and maintaining flumes constructed from either concrete or rock (Figure 4.4).

Figure 4.2 *Plugged Deck Drain.*

Figure 4.3 *Flush Truck in Operation.*

Figure 4.4 *Rock Flume-Off of Bridge Approach.*

Whether off of the deck or through the deck, drainage should be directed to designated locations to ensure that it contributes to neither erosion at the berm nor to deterioration at the bearings and substructure. Drainage

Figure 4.5 *Drain Extension Retrofit.*

through the deck should be conveyed below the top of the bearing seats (Figure 4.5).

4.4.2 Deck Patching

The second priority in deck maintenance is maintaining a relatively smooth wearing surface. For concrete decks, spalling may occur at the edges around expansion joints and drains, or anywhere that corrosion of the underlying reinforcing steel occurs. The spall results in edges that unravel further under subsequent traffic and with freeze/thaw cycles, resulting in growth of the deteriorated area. The best course of action is to address this by patching the area – removing slightly past the damaged area and replacing with new concrete or other patching material. Although maintenance crews often carry out small patching operations of a few square yards, large patching operations are typically handled through substantial maintenance projects. Concrete patching is discussed in Chapter 5 (Figures 4.6 and 4.7).

The patching of a small portion of bituminous wearing surface is typically done with cold mix asphalt. Maintenance crews typically are experienced

Figure 4.6 *Deck Deterioration at Expansion Joint.*

with this from work on asphalt roads. If there is extensive patching required on either a concrete or a bitumunous wearing surface, an overlay of the deck should be considered. This requires extensive removal and preparation of the existing wearing surface and is considered substantial maintenance and is discussed in Chapter 5.

Deteriorated timber deck planks should be replaced if they are resulting in a rough ride for traffic. In areas where planks have bowed or cupped, resulting in a ponding of water but not yet compromising the flow of traffic,

Figure 4.7 *Area Prepared for Partial Depth Patching.*

small drains may be installed through the planks to prevent unnecessary water damage (Figure 4.8).

Decks made from FRP are a relatively recent addition to bridge inventories and are currently few in number. Due to the proprietary nature of these products, the manufacturers' recommendations should be consulted for appriopriate maintenance activities.

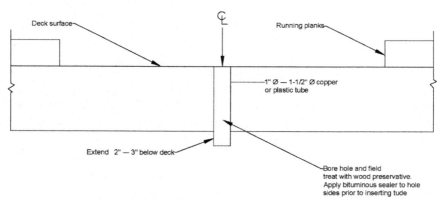

Figure 4.8 *Drain Tube Through Timber Planks. (Adapted from FHWA BM [3]).*

4.4.3 Deck and Crack Sealing

The third priority in deck maintenance is sealing or waterproofing the material. Both concrete and bituminous wearing surfaces have cracks. A crack in the surface of the material allows the ingress of water. For both materials, freezing of that water may result in an expansion of the crack. In a concrete deck, once a crack extends to the top layer of reinforcing steel, corrosion of that steel may occur. A crack width of 0.18 mm is sufficient to allow the ingress of chlorides and moisture [4] (Figure 4.9).

Cracks in bituminous wearing surfaces may be sealed (for cracks <0.75 in. wide) or filled (>0.75 in. wide) with thermoplastic material the same as with bituminous pavements. Care must be taken, however, to not damage any waterproofing membrane under the wearing surface. Before beginning any extensive preservation action on a bituminous wearing surface on a bridge, consideration should be given to the expected life of such overlays. A bituminous overlay and membrane placed on an existing structure whose condition indicates that it is near the end of its service life may have been placed with the expectation of a 5-year life; while such a system placed on a new structure may be intended to last 20 years.

Cracks in concrete surfaces may be sealed with one of a number of different types of materials. Table 4.1, adapted from the FHWA *Bridge Maintenance Training Reference Manual* [3] shows properties for the various materials discussed further. The selection of material first depends on whether the intent is to seal individual cracks (crack sealing), or to waterproof the entirety of the wearing surface (deck sealing). Deck sealers create a barrier to penetration by water and accompanying chlorides. They may be categorized by depth of penetration into the pours of the concrete. Silane and siloxane sealers create a nonwetting (hydrophobic) barrier by reacting with

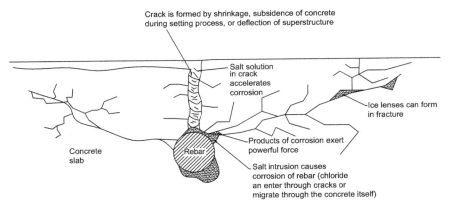

Figure 4.9 *Spall Development From Cracking (AASHTO Maintenance Manual [4].)*

Table 4.1 Relative properties of concrete sealers

Sealers, coatings, and membranes

Material	Thickness				Properties											
	Penetrating sealer	Surface sealer 0–30 mils	High build coating 10–30 mils	Membrane 30+ mils	Bond	Abrasion resistance	Chemical resistance	Elongation	Chloride ion resistance	Water absorption resistance	Vapor transmission	Penetration	Ultraviolet resistance	Life expectancy	Water resistance	Cost
Linseed oil	X								P	P	P	P	P	L	G	Low
HMWM	X								G	E	F	E	V	E	P	High
Silicate	X								F	P	G	F	F	G	G	Low
Siloxane	X								E	E	V	G	G	G	V	Mid
Silane	X								V	V	E	E	E	E	E	High
Latex		X	X		G	F	P	–	Ratings							
Silicone		X			F	P	P	–	E – Excellent							
									V – Very good							
Methacrylate		X			G	G	F	–	G – Good							
Epoxy		X	X		E	G	F	–	F – Fair							
									L – Low							
"Hard" urethane		X	X		F	E	G	–	P – Poor							
Elastomeric Urethane				X	F	F	P	E								
Vinylester				X	E	G	E	–								
Polyester				X	G	G	E	–								

Source: Adapted from FHWA Bridge Maintenance Manual [3]

the cement paste inside of the pours. Epoxies and linseed oil form a film at the surface of the pours to prevent penetration (Figure 4.10).

Crack sealers may be applied to individual cracks or across the entirety of the deck surface in a flood coat. Crack sealers to bond with sufficient strength to restore some structural capacity to the material. The most commonly used crack sealers are epoxies and high molecular weight methacrylate (HMWM). Methacrylate and polyurethane are also used as crack sealers.

Figure 4.10 *Deck Sealing Operation.*

A 2009 study by the Minnesota Department of Transportation comparing research and the experiences of other DOTs with various sealers found silane to outperform either siloxane, epoxy, or linseed oil as a deck sealer. For crack sealers, either HMWM or epoxy sealers could be effective. HMWM was applicable for decks with extensive cracking, due to the flood coat application, for decks with fine cracks due to its very low viscosity. Epoxy sealers were applicable for decks with few cracks because they are typically applied to individual cracks, and for larger cracks due to its higher bond strengths [5].

Various owners and agencies have different opinions as to the sealer material of preference. This is due to variations in cost, surface preparation, and environmental application requirements between the material as well as the individual agency's previous experience with the products. In general, all of the sealers require cleaning of the concrete by sandblasting, then air cleaning, and are applicable only on a dry substrate and only over a range of relatively mild ambient temperatures.

4.4.4 Deck and Expansion Joint Washing

A 2013 study by the Washington State Department of Transportation found that half of 34 states responding to a survey of practice pressure-washed bridge decks on a regular basis, typically annually or biannually. The washing of bridge decks was a common practice among those states in the Northeast and Upper Midwest [6]. These states receive more snow relative to others in the United States, resulting in greater usage of deicing salts on the roadways. These states also tend to have an older inventory of bridges with steel superstructures and bearings devices. Research by the Oregon Department of Transportation found that washing of concrete bridge decks at weekly or greater intervals does not reduce the concentration of chlorides on the surface of the deck; however, it does appear to slightly reduce the absorption of chloride ions into the concrete [7]. The Washington State report found some states which had a program of washing bridge decks in years prior to the report had discontinued their programs because of costs to meet environmental regulations requiring the collection and treatment of effluent from washing.

The washing of expansion joints and bearing seats was a more common practice. Twenty-five of the responding states in the report regularly washed either or both of these areas. A higher rate of practice might be expected for a number of reasons. First, runoff from spring and summer precipitation might be expected to flush some lingering chlorides from the top of bridge decks, while bearing seats are shielded from the direct flow of runoff. Second, debris may impede the proper functioning of these devices. Lastly, bearing seats provide a convenient location for roosting birds (most notoriously, pigeons) whose droppings facilitate corrosion of common construction materials.

4.4.5 Timber Deck Preservation

Timber decking may suffer from the growth of fungi or algae in the presence of available water. Even pressure-treated timber may begin to degrade after 5 years of exposure if not sealed [3]. Timber decking should be pressure washed and treated with a preservative containing either pentachlorophenol or copper naphthenate every 3–5 years.

4.5 BRIDGE SUPERSTRUCTURE AND SUBSTRUCTURE

4.5.1 Washing Superstructures

In another 2013 report, the Washington State Department of Transportation surveyed states as to their practice in washing steel superstructures [8]. Problems in attributing the origin of a high percentage of the responses received precluded any definite conclusions on nationwide practice, but it could be ascertained that washing steel superstructures on an annual or biannual basis was a common practice in the northeast United States and far less so in other parts of the country. Although it was assumed that washing would prolong the life of the structural steel and its protective coatings, the report found that no research providing definite correlations had been conducted and, so, proposed conducting such research in Washington State.

It was found that the Rhode Island Department of Transportation had conducted an analysis in 2002 comparing the costs of their bridge washing activities on 45 Interstate steel bridges with good paint condition at the beginning of the study period to an alternative of no washing; based on assumed transition probabilities to the paint condition ratings in their

PONTIS Bridge Management system. The DOT found that washing would save $882,000 over an 8-year period [9].

4.5.2 Sealing Bearing Seats

Washing of the bearing seats is an activity that would typically be performed in conjunction with washing the deck and cleaning out expansion joints, as previously discussed. Similar to concrete decks, the concrete surfaces at the bearing seat, including adjacent concrete backwalls and girder ends, may be sealed to prevent water intrusion. A number of the same products are applicable here, as on the deck. At the bearing seat, though, wear from traffic is not a concern. What should be avoided at these locations is applying a membrane-type sealant that does not penetrate into the concrete substrate. A membrane-type sealant forms a relatively thick intact barrier to water. A membrane material which does not seep into the pours of the underlying concrete may also act to trap water at the interface between the material if holidays (i.e., tears) appear in the membrane (Figure 4.11).

Sealing concrete in the bearing seat area is a good practice, particularly if reinforcing steel in the adjacent area lacks any other protective system (such

Figure 4.11 *Failed Waterproofing Member and Bearing Area Deterioration.*

as epoxy coating) against corrosion. The entirety of the horizontal surface should be seated, as should adjacent concrete surfaces in the splash zone. This would typically include the vertical surface of any adjacent walls and 2–5 ft. of adjacent concrete beams or girders.

4.5.3 Bearing Device Maintenance

Other than keeping bearing devices clean, the only preventative maintenance actions that may apply to them are lubrication of metal bearings on the surfaces around the pin, and reinserting any teflon sheets that may have slipped out of a polytetrafluoroethylene bearing. Both of these actions are to ensure that the bearing can move as intended and both typically require jacking the superstructure load off of the bearing to provide access to the interior surfaces of the device. Modern bridge designs typically include detailing that will allow placement of a jack and access to the bearing device to provide for either servicing or replacing it. See the discussion in Chapter 5 on bearing device replacement for discussion of jacking.

4.5.4 Painting

Maintaining a functional paint system on a steel bridge protects it from corrosion, preventing deterioration. This is exactly the goal of preventative maintenance actions; however, due to the high cost of painting bridges, bridge paint projects are not generally considered as such. The full repainting of an existing bridge requires: providing access; establishing traffic control; removing, containing, and disposing of the existing paint system; surface preparation; and, finally, applying the new multipart paint system. Typically, bridge painting is included in the purview of substantial maintenance work.

If because of the cost, the paint system is considered as a distinct element of the bridge, spot painting can be considered as preventative maintenance. Spot painting is painting of only the sections of the bridge where the existing paint system is compromised. The *AASHTO Maintenance Manual for Roadways and Bridges* suggests that spot painting may be a viable option for structures with up to 35% of the painted area compromised [4]. As with a full repainting of a bridge, the general steps are to clean the area of concern of surface contaminants such as dirt and grease, prepare the surface for the adhesion of the new paint, and apply the paint system. Much of the effort in this process is involved in the surface preparation. The complete repainting of a bridge would require removal of the existing paint down to the bare metal, and cleaning of the bare metal to a "near-white" condition. For the spot-painting procedure discussed later in the chapter, not only is the area of preparation greatly minimized, but also the degree of preparation as well.

If the existing paint system contained lead, as was common prior to 1980, both worker and environmental protection measures will have to be taken if any of the existing paint is removed. These measures include worker protection (protection from inhalation or skin exposure and testing for levels of lead in the blood), containment (with tenting and the use of negative pressure air flow), soil testing of adjacent environs for inadvertant contamination, and disposal of the wastings at a suitable facility.

Note, that although the bridge community in the United States ended the use of lead-based paint systems in the late 1970s, many of the vinyl paint systems used immediately after had sufficient quantities of lead in their formulation to require treatment as lead-based paint. If removing any paint, it is recommended to have a Toxicity Characteristic Leaching Procedure test done on the existing system. This is a test that simulates the leaching that may occur to waste disposed off in a sanitary landfill, and therefore determines how the waste must be dealt with by Environmental Protection Agency regulations in the United States. If the existing lead paint is left in place, a great deal of cost may be avoided. If the existing paint produces a lead content higher then 5 mg/L, it must be treated as a hazardous waste [10].

To avoid this expense, many agencies use overcoating for spot-painting activities. Overcoating involves cleaning, removing loose existing paint and then applying the new system over the mixed substrate, which may include existing paint, bare steel, and corroded steel. Overcoat systems are "surface tolerant," they adhere over a range of substrate conditions. Cleaning is usually performed by power washing. The surface preparation of removing loose existing paint may be done with hand tools or vacuum shrouded power tools. Wastings of lead-based paint must still be treated as hazardous, but are much less in volume. Also, the elimination of a need to sand blast a large area for surface preparation generally relieves the need to enclose the location to capture airborne particulates. Cost may be reduced 35–75% compared to traditional removal and paint methods [4].

There are multicoat and single-coat paint systems available. Testing by the FHWA found the best performance for overcoating were provided by three-coat moisture-cured urethane systems and three-coat epoxy-based systems. These systems have a prime, intermediate, and finish coat [11]. For spot painting by maintenance personnel, one-coat systems can provide acceptable performance for considerabily less expenditures in resources and effort. The best performing one-coat system was the calcium–sulfonate alkyd resin system.

4.6 BRIDGE SUBSTRUCTURE AND WATERWAY

Piers adjacent to roadways under bridges are subject not only to direct impacts from errant traffic, but may be exposed to deicing salts sprayed off the road from the tires of passing vehicles. The columns and stems of such piers in a splash may benefit from application of concrete sealers.

Piers adjacent to waterway channels face a different set of challenges. After high water flows, maintenance crew may find debris in the form of drift wedged between column, or even up into the superstructure. It is essential to remove this in that remaining drift may serve to catch more debris in subsequent high water flows. This serves not only to increase the hydraulic force on the piers, but by narrowing the available waterway opening, backwater elevations are increased as is the velocity of the flow under the bridge. Even after the water levels go down, drift presents a possibility for fires to start under the structure (Figure 4.12).

High-water velocities can result in scouring away of the streambed material from foundation elements. Even if the foundation elements are founded deep enough into the geological strata to provide safe capacity, the exposure of an element such as a steel pile to cycles of wet and dry

Figure 4.12 *Example of Drift Accumulation.*

Figure 4.13 *Scour at Pier Wall.*

will encourage corrosion. Maintenance crews can protect scoured locations from further deterioration by filling in with large diameter rock. The
presence of the rock will also encourage the redeposition of sediment at
the location. The size of rock depends on the expected water velocity it is
expected to resist. In lieu of an engineered analysis, the FHWA *Maintenance
Training Manual* suggests that stones of a minimum size of 50 pounds to
be employed [3]. Bags of concrete, concrete blocks, or proprietary concrete shapes (e.g., "A"-jacks) might also be used. Underwater placement of
poured concrete with a tremie is typically beyond the scope of preventative
maintenance (Figures 4.13 and 4.14).

Erosion due to drainage from the roadway can also be addressed with
proper grading and the placement of rock. Rock or concrete placed to protect a berm is also known as revetment. Left unchecked, improper drainage
at the berms can find its way back to the abutment and expose the foundation elements under the abutment beam (Figures 4.15 and 4.16).

4.7 APPROACHES AND ROADWAYS

4.7.1 Driver Guidance

Highway bridges exist to convey vehicular traffic. However, errant traffic
represents a threat to every highway bridge. Drivers departing from their

Figure 4.14 *Maintenance Forces Placing "A-Jacks."*

lanes may strike bridge railings on the deck and impact piers at grade-separated structures. Overheight vehicles may impact superstructures above at-grade separations and structures such as through trusses. Overweight vehicles may procede over a bridge without obvious effect – unless they are grossly overweight for the structure – but the cumulative effect of regular passage of overweight traffic will be to shorten the service life of a bridge. The role of preventative maintenance in all of these scenarios is to provide proper guidance to drivers.

Lane departures can be minimized by clearing delinating lanes with striping and reflective guideposts. This is of particular importance in that drivers have a tendency to laterally reposition toward the center of the

Figure 4.15 *Erosion Under Abutment.*

roadway and away from the edge of a bridge [12]. Object markers serve to remind a driver that he or she is passing over a structure, with its attendant appurances in the roadside environment. In the case of a buried structure, the markers can serve to notify the possibility of a dropoff on the sideslopes that might not be obvious from the roadway at speed.

Figure 4.16 *Use of Stone as Revetment at Berms and Scour Protection at Piers.*

Keeping the roadside well graded and free of an overgrowth of vegetation facilitates both drainage and a clear view. An analysis of crashes at bridges on the Kansas state highway system found a negative correlation between crash rates and substantial maintenance work on the bridge deck in prior years [13]. This work not only improves the quality of the ride on the bridge, but is typically by improvements in the approach roadway such as clearing, grading, and striping.

Proper advanced signage alerts drivers to weight and clearance restrictions at a bridge site. It is important that maintenance personnel maintain not only physical signage, but that they ensure any central database for structures on their highway system is kept current. Permits for overweight, overwidth, and overheight vehicles are typically issued by a central office in each jurisdiction. When issuing route specific permits, routes are vetted against the available information.

4.7.2 Approach Settlement

The most ubiquitous problem concerning approaches and bridge maintenance is the "bump at the end of the bridge", typically resulting from settlement of the approach relative to the bridge. Bridges are usually founded on deep foundations to ensure that they will not settle, while approach roadways are treated as pavement structures founded on the soil mantle at the site. Proper drainage away from the roadbed will do much to prevent settlement of the approach. Settlement is often caused as the fine material in the underlying soil matrix is washed out, leaving a void underneath the approach (Figure 4.17).

Figure 4.17 *Settlement of Approach Slab.*

There are three options to alleviate this situation:

1. Form a wedge in the wearing surface to transition from the approach to the deck;
2. Raise the approach roadway; and
3. Replace the approach roadway.

If the difference between the deck and the approach is only a fraction of an inch, a transition can be formed from a wedge of material (asphalt for a bitmunious roadway, polymer concrete for a reinforced concrete roadway). A gentle transition of at least 1:50 (rise:run) should be formed.

If the approaching roadway consists of concrete pavement, it may be raised by filling the underlying voids with either a cementitious grout (mudjacking) or a high-density polyurethane (foamjacking). The process is the same in either case, only the material used is different. The process may also be referred to as concrete leveling or concrete raising. The operation consists of drilling a series of regularly-spaced holes, temporarily plugging all but two and filling the underlying void through one hole until the jacking material is visible at the other hole; then alternately opening and pumping into other holes to raise the pavement into place.

Foamjacking uses smaller equipment, is relatively quick, results in a waterproof mass under the raised area, and does not contribute additional weight that may overburden the underlying soil. Mudjacking is typically less expensive and it is more likely that maintenance crew will have the required equipment and experience for operation with grout (Figures 4.18 and 4.19).

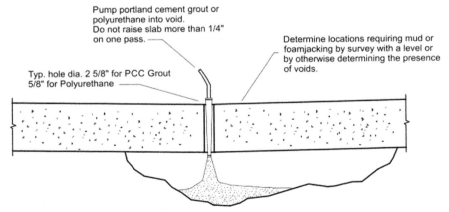

Figure 4.18 *Pumping Grout Into Underlying Void.*

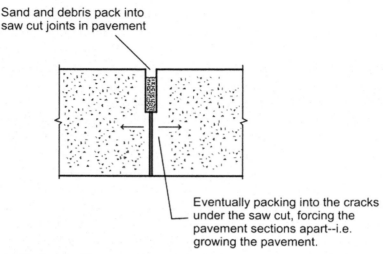

Sand and debris pack into saw cut joints in pavement

Eventually packing into the cracks under the saw cut, forcing the pavement sections apart--i.e. growing the pavement.

Figure 4.19 *Concrete Pavement Expansion.*

4.7.3 Relief Joints

Approaches can also suffer distress from the growth of adjacent concrete pavement. Concrete pavement grows in length as sand or other incompressible particles penetrate into the saw cut contraction joints in concrete pavement over cycles of cold weather contraction and placement of salt and sand during the winter. To compensate, maintenance forces can cut relief joints at the road-end of the approach and install a compressible watertight joint. An opening, no more than 4 in. in width is cut. A number of joint systems are available, but a polymer impregnated polyurethane foam system appears to be the easiest for maintenance crews to install and maintain (Figure 4.20).

4.8 RECOMMENDATIONS

Preventative maintenance is intended to preserve the good condition of the elements of a bridge rather than to restore damaged element to their original state. These actions have the potential to save bridge owners' significant funds in construction and maintenance actions and to keep facilities in good operation with the least impact to users. Though the individual actions are relatively cheap, they must be applied consistently from early in the life of the facility for best impact.

Figure 4.20 *Installation of Pressure Relief Joint.*

REFERENCES

[1] Kong JS, Frangopol F, Gharaibeh E. Life prediction of highway bridges with or without preventative maintenance. Proceedings of the eighth ASCE special conference on probabilistic mechanics and structural reliability; 2000. pp. 24–26.

[2] Yanev B. The management of bridges in New York City. Eng. Struct. 1998;20(11): 1020–6.

[3] Federal Highway Administration. Bridge maintenance training. Washington, DC: US Department of Transportation; 2003.

[4] AASHTO Highway Subcommittee on Maintenance. AASHTO maintenance manual for roadways and bridges. Washington, DC: American Association of State Highway and Transportation Officials; 2007.

[5] Johnson K, et al. Crack and concrete deck sealant performance. St. Paul, MN: Minnesota Department of Transportation; 2009.

[6] Berman, JW, Roeder CW, Burgdorfer R. Standard practice for washing and cleaning concrete bridge decks and substructure bearing seats including bridge bearings and expansion joints to prevent structural deterioration. Olympia, WA : Washington State Department of Transportation; 2013. p. 201.

[7] Soltesz S. Washing bridges to reduce chloride. Salem, OR: Oregon Department of Transportation; 2005.

[8] Berman J, Roeder C, Burgdorfer R. Determining the cost/benefit of routine maintenance cleaning on steel bridges to prevent structural deterioration. Olympia, WA: Washington State Department of Transportation; 2013.

[9] Boardman J, Brownell J, Nickelson J. Bridge inspection/washing program. Providence, RI: Rhode Island Department of Transportation, Operations Division; 2002.

[10] Test Methods for Evaluating Solid Waste, Physical/Chemical Methods. [Online] Environmental Protection Agency, February 2007. Available from: http://www.epa.gov/solidwaste/hazard/testmethods/sw846/index.htm. SW-846.

[11] FHWA Bridge Coatings Technical Note: Overcoating (Maintenance Painting). Technical Note. [Online] Federal Highway Administration, January 1997. Available from: http://www.fhwa.dot.gov/publications/research/infrastructure/structures/bridge/overct.cfm

[12] Ogden KW. Crashes at bridge and culverts. Melbourne, Victoria: Monash University; 1989. Report No. 5.

[13] Hurt M, Schrock S, Rescot R. Review of crashes at bridges in Kansas. In: Proceedings of the 2009 mid-continent transportation research symposium; 2009. Ames, IA: Iowa State University.

CHAPTER 5

Substantial Maintenance and Rehabilitation

Overview

Substantial maintenance actions are discussed in this chapter. Considerations for determining scope of work and the level of repair are examined. The repair methods and considerations specific to concrete and steel are reviewed. Common substantial maintenance activities specific to particular bridge elements are discussed.

5.1 INTRODUCTION

Maintenance can be defined as activities that are undertaken to mitigate the deterioration of an item or system and to maintain its function. Implicit in this definition is the assumption that the performance of an item or a system will degrade over time. Typically this degradation is in proportion to the use and exposure it receives over that time. Highway bridges deteriorate in proportion to the number and weight of the vehicles that pass over them and in proportion with their exposure to deleterious items in the environment, such as road salts. An individual bridge, however, does not deteriorate as a monolithic unit. The individual elements of each bridge have a different exposure to traffic loads and to the environment. The condition of an individual element may, and often does, deteriorate to an unacceptable level while the rest of the bridge is in good shape.

The owners of highway bridges face the challenge of providing structures that perform their function – to carry vehicular traffic – at an acceptable level of service and at low cost. It is not only costly to replace bridges, but waiting for deterioration to proceed to such a level as to justify the incursion of the cost will entail a period of low level of service from the structure. For a highway bridge this usually takes the form of restricting loads and traffic speeds on the structure. Typically an owner will also incur increased costs from increased frequency in inspection and in low-level maintenance tasks, such as filling potholes, required just to keep daily operation. The most cost-effective approach is to repair deteriorated elements and to maintain the bridge as a whole at an acceptable level of service.

Highway Bridge Maintenance Planning and Scheduling
http://dx.doi.org/10.1016/B978-0-12-802069-2.00005-2

Substantial maintenance actions restore individual bridge element to a desired condition. Typically this is as close as possible to their original condition, but a lesser condition may be acceptable if it is consistent with the goal of providing a bridge with an acceptable level of service. Whereas preventative maintenance actions are intended to slow the rate of deterioration, substantial maintenance actions repair the deterioration. The maintenance is substantial in effect and in cost. Typically the work is performed under contract with highway construction contractors who specialize in such work.

If the intent is to address only the most pronounced deficiencies, substantial maintenance work may be conducted on a particular bridge element, while other compromised elements are left as is. If sufficient work is conducted to bring the entire bridge back to its original or better condition, this activity is considered rehabilitation. Rehabilitation will also typically be conducted with the intent to improve the functional characteristics of the bridge (such as roadway width) to meet current specifications.

5.2 ASSESSMENT AND SCOPING

Determining the scope of substantial maintenance work to be conducted on any particular bridge requires first confirming that repair is the best option. Closing a bridge to service may be the most cost-effective solution. If it is determined to repair the bridge, then the level and extent of the repair must be selected. The level of the repair effort can be anywhere on a continuum of scope, from minimal (conducting only enough work on specific elements to keep the bridge at a minimally acceptable level of service for a short period) up to a level of improving the capacity of the structure beyond the original design to meet current criteria.

5.2.1 Closing or Removing Bridges

When assessing a bridge for substantial maintenance work, one of the first tasks is to check to see if that bridge is scheduled for removal or for extensive rehabilitation within a few years. Highway facilities have their own maintenance needs, and changing traffic patterns and volumes may require reconstruction of those facilities to meet current needs. Changes in roadway widths and alignments may require reconstruction of bridges at that time, limiting the period to recover value from the investment of substantial maintenance work. Even if the existing bridge is to remain in place as part of future highway work, waiting to conduct substantial maintenance actions at that time can take advantage of any traffic control measures in place for that project. This saves cost and minimizes inconvenience to users.

If no future work is planned at the site, however, closing the bridge or even physically removing the structure may be the preferable course of action if one of the following three questions may be answered in the affirmative:

1. Would it cost less, including user impacts, to replace the bridge rather than repair the deficiency?
2. Is the bridge crossing still required?
3. Would a lesser level of service be more appropriate for the site for current traffic volumes and characteristics?

In the case of extensive deterioration to the superstructure of some structure types – such as reinforced concrete girders integral with the deck – repairs may require significant shoring and construction phasing. This could result in long interruptions to traffic and a high construction cost for the repair. If repair work would cost a significant portion of what full replacement would cost, and if the functionality and/or future life of the end-product would be less than that of a new bridge, replacement may be the preferred option.

Traffic patterns are not static. Populations, industry, and commercial activity may move away from an area, resulting in reductions in traffic on once busy routes. If a bridge serves primarily to facilitate access to a particular area and there has been a significant decline in volumes utilizing the bridge, and if access might be sufficiently served by other existing routes, then removing the bridge should be considered. If the bridge is servicing a low-volume through route, or if a lower level of service for access to an area is more appropriate, the bridge might be removed and replaced with either a lower water crossing (for channel crossings) or an at-grade intersection (for overpass structures).

5.2.2 Level of Repair

When assessing deterioration, the actual impact of the deterioration on the functioning of the structure should be considered. If the deterioration is in a low stress area and does not significantly impede the capacity of the bridge, it may be appropriate to limit the scope of work to alleviating the source of corrosion and protecting the area from further corrosion.

If deficiencies are more severe, yet the bridge is scheduled for replacement within a few years, or if a bridge restored to full condition would still not meet desired function characteristics (e.g., a functionally obsolete bridge), the proper level of repair may be to limit work to maintain it in operation for only a few years. As an example, an appropriate scope of repair for a bridge with significant deck deterioration which will be replaced in 5 years during a programmed highway reconstruction project might be to place a waterproof membrane and bituminous overlay rather than concrete deck patching and rigid concrete overlay. The former work might have an expected life of less than 10 years, yet will last the remaining life of the bridge and will be cheaper in terms of construction cost and user impacts.

A scope of repair that results with element conditions for a bridge that are less than those as originally constructed may be selected if the degree of disruption to users required for work to restore to full condition is not tolerable. The extent of repair may be limited to that what can be done in a limited time (e.g., overnight or over a weekend). Less than original condition might also be avoidable if the deterioration of compromised elements or their supporting elements are such that return to the full original condition is not possible.

While the original level of condition is usually the goal of substantial maintenance work, sometimes the original level of performance is deficient compared to current demands. If the deficit is specific and remediable, a retrofit may be done to improve the performance of particular elements. For example, if the girders of a bridge lack the capacity for extralegal truck traffic (i.e., overweight trucks), but the deck and substructure elements have sufficient capacity, then there are methods of adding capacity (e.g., FRP wrap) to strengthen the girders and improve the overall load-carrying capacity of the bridge.

5.2.3 Design Codes and Specifications

To properly design substantial maintenance repairs, a designer should be familiar with the previous design codes and specifications used for the bridges at hand to understand what assumptions were made in their design process. Design codes and specifications evolve over time as experience is gained, the weights of trucks increase, and design tools and methods become more sophisticated. Previous versions of design codes may have specific deficits (e.g., an overestimation of the shear capacity of concrete in the American Association of State Highway Officials (AASHO) code of the 1950s) that manifested in particular defects, which may be expected in bridges of a particular vintage.

Inherent in all design codes are assumptions about the physical behavior and responses of structures. These concepts are assumed to simplify the analysis and design process and are intended to be conservative in nature. That is, any inaccuracy in calculations of capacity should result in underestimates and those for response should result in overestimates. Repair work deals with the results of the *actual* structural responses. Sometimes a design which seems conservative may be the source of a problematic structural response.

As an example, in the 1970s, bridge designers did not weld connection stiffeners (stiffeners connected to cross-frames or to diaphragms) to the tension flanges of steel girders. Designers knew that fatigue cracking could initiate at the toes of welds in tension areas due to the effects of restraint and residual stresses. Years later, as these bridges had experienced many cycles of loading, fatigue cracks began appearing in the girder webs at these locations. This was a result of out-of-plane distortion (discussed in Section 5.3.2.2). In the mid–1980s, standard practice changed to requiring a positive attachment of the connection stiffeners to both flanges. The best available knowledge was reflected in the design codes and practices of the time, yet there were intended consequences.

In addition to being aware of limitations in previous design codes and practices, a designer should be cognizant of the variance in proportioning members that may be expected between current and previous codes. Current code may indicate that a smaller member will satisfy the capacity requirements for a certain loading condition than previous codes had. Using a smaller member may satisfy the current code, but may result in an unintended change in the response of the structure as a whole.

5.3 REPAIR METHODS

The following section discusses general repair methods specific to the material repaired. Typical applications of these methods to common repairs for specific elements are discussed in subsequent sections.

5.3.1 Concrete

5.3.1.1 General Discussion

The most commonly used material in modern highway bridge construction is concrete. Because of its relatively low-tensile capacity it is almost always combined with bars of a high-tensile capacity material, such as steel or glass, to form a composite material, reinforced concrete. The reinforcing may be placed in the concrete member in an unstressed state, or the reinforcing material may be stressed in tension to induce compressive stresses in the concrete section. These stresses are induced in locations in a structural

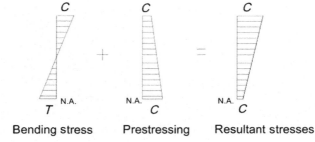

Figure 5.1 *Effect of Tensioning on Beam Stresses.*

element where its dead load will induce tensile stress, effectively cancelling them. If the reinforcing is tensioned prior to curing of the concrete, the member is "prestressed". If tensioning is applied to a hardened concrete member, the member is "post-tensioned" (Figure 5.1).

Even bridges that utilize steel superstructures typically have decks, piers, and abutments formed with reinforced concrete. The majority of substantial maintenance activities on highway bridges consists of repairs to concrete members.

Because of the low-tensile strength of concrete almost all damage to the material, whether from deterioration or from an external cause such as an impact, has essentially the same manifestations. The concrete will crack as the tensile capacity of the concrete matrix of cement paste and aggregate is exceeded. In a reinforced member, as the cracking grows, the bond between the reinforcing steel and the concrete is compromised and concrete begins to debond, or delaminate, from the reinforcing steel. As the cracking further progresses the concrete fractures through and the damaged section spalls off. Once cracking begins, water and contaminants such as chlorides may enter and begin the corrosion process with steel reinforcing.

5.3.1.2 Crack Repair

Repairs early in this process are directed to sealing cracks. All concrete members are subject to some degree of cracking initiated by shrinkage due to loss of water in the initial curing process and which may continue throughout the life of the member as it is subject to thermal expansion and contraction and to the deflections inherent in the service of any structural member. As discussed in Chapter 4, concrete sealers may be applied as a preventative maintenance action to keep water from penetrating into the member. Gravity poured sealers are suitable for waterproofing cracks of a width up to 0.08 in. (2 mm) [1]. Wider cracks can be treated by routing and

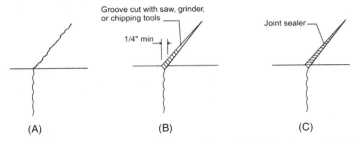

Figure 5.2 *Crack Treated by Routing and Sealing.* (A) Original crack. (B) Routing. (C) Sealing.

sealing as shown in the Figure 5.2, but more commonly wider cracks are treated by epoxy injection.

Selection of the proper epoxy material requires deciding whether to restore the structural capacity of the damaged section of concrete or whether to simply seal the crack. To restore capacity, a structural epoxy is used, which rigidly bonds the adjacent concrete surfaces. If restoring capacity is a primary consideration for treating the crack, epoxy injection may be used for cracks as narrow as 0.002 in. (0.05 mm) (Figure 5.3). If a crack is not dormant (i.e., it cycles through closing and opening, as might be caused by thermal movements), is not growing in length, and does not extend deeply

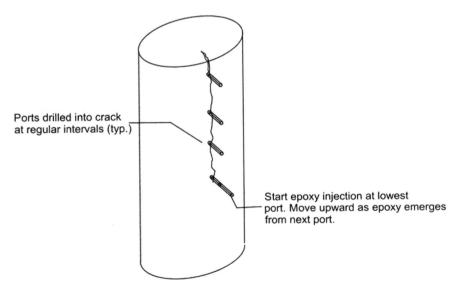

Figure 5.3 *Epoxy Injection.*

enough to impair the structural capacity of the member; then it may be sealed with a more malleable nonstructural epoxy.

Whether treating a crack by routing and sealing, or by epoxy injection, preparation is important. Cracks should be cleaned and dried according to the recommendations of the manufacturer of the sealing material. Further preparation for epoxy injection requires drilling holes for porting the epoxy into the substrate under pressure.

5.3.1.3 Patching and Surface Repair

Once concrete debonds from the underlying reinforcing steel and begins to fracture along a plane parallel to the surface, it is delaminating. Areas of delamination will eventually spall off, exposing the interior of the concrete member and reinforcing, resulting in further deterioration. Alternately, an impacting force may fracture the concrete, resulting in its debonding and falling away. In either case, damaged concrete must be removed and replaced with new material. This is referred to as concrete surface repair, or more vernacularly as patching.

Concrete patching is categorized by how far it extends into the thickness of the section of the concrete member. The AASHTO *Road and Bridge Maintenance Manual* categorizes patching as either Type A, B, or C, where:
- Type A extends only through the cover, that is, to the outside surface of the outer layer of reinforcing steel.
- Type B extends from the surface to at least 1 in. behind the outside mat of reinforcing steel.
- Type C extends completely through the member.

Types A and B patching are also commonly referred to as partial–depth patching; while Type C is referred to as full-depth patching.

Large areas of Type A patching are not practicable on either vertical or overhead patching as that the weight of the patch material must be completely borne by the adhesion of patch material to the existing material. Type B is the most common type of patching. The gap below the exterior layer of reinforcing is to allow the flow of patch material behind the reinforcing to anchor the patch. The depth of the gap may be adjusted as required to allow aggregate in the patch material to pass through. Type C patching presents not only the challenge placing and consolidating a greater volume of patch material, but with forming the backside of the patch (Figure 5.4).

Once damaged concrete is removed, if it is discovered that over 20% of the original cross-section of an exposed steel reinforcing bar is lost, then additional reinforcing should be spliced to the original bar. For circular

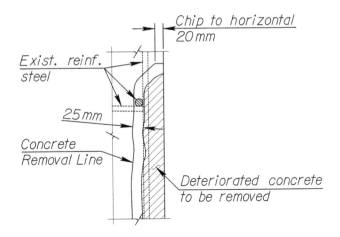

TYPICAL CONCRETE REMOVAL DETAIL

CONCRETE SURFACE REPAIR: Remove loose, cracked, and delaminated concrete as directed by the engineer. Repair the area(s) with Shotpatch 21F shotcrete or an approved equivalent as outlined in the Special Provision. Sandblast and prepare areas to be repaired prior to placing shotcrete. Repair or replace deteriorated reinforcing steel at the direction of the engineer. Place shotcrete to match existing surfaces. On overhead surfaces provide a minimum of 25 mm cover over the reinforcing steel.

Figure 5.4 *Typical Concrete Patching Details.*

reinforcing bars, a 20% reduction in area corresponds to a slightly more than 10% reduction in diameter. For smaller bars (#4 and #5), as typically used in decks and walls, usually the spliced bar matches the size of the existing and a class B lap splice length is assumed. For larger bars, as common in pier stems and columns, an engineer should determine the additional reinforcement needed to be spliced, as that the lap splice length of larger bars is significantly longer than it is for the smaller bars. Alternately, mechanical rebar connectors may be used develop reinforcing in a shorter length.

5.3.1.4 Material Selection

Selection of patching material requires consideration of four factors: method of application, strength, speed of cure, and corrosion resistance. Concrete surfaces may be repaired in either horizontal, vertical, or overhead orientations. Material for a horizontal patch may be readily poured into place. Material for either a vertical or overhead patch may also be poured into place with the use of formwork to retain the material. The same regular Portland Cement Concrete used in new construction may also be used for a poured patch.

If it is not possible, or desired, to place formwork, the patching material may be pneumatically applied, that is, sprayed into place with air pressure. The common term for this material is shotcrete. Shotcrete is a mortar of cement paste, fine aggregate, and water. If the mortar is hydrated prior to being pumped, the shotcrete is termed a wet mix. If it is hydrated at the nozzle as it is being sprayed, it is a dry mix.

There are many proprietary "prebagged" concrete and mortar products available for use as patching materials, and many admixtures available for mixed concrete, that confer properties such as higher compressive strength, higher early strength, decreased permeability, and corrosion resistance. An important characteristic for many patching applications that is overlooked when specifying the material is shrinkage. Cementitious materials tend to shrink during the curing process. Shrinkage may result in cracking of the patch material. This may be compensated for by the additional of expansive material to the formulation (shrinkage compensated concrete) or by the addition of fibers to resist the tensile stresses of shrinkage.

5.3.1.5 Corrosion Resistance

The corrosion of reinforcing steel in concrete is an electrochemical process. In chloride-contaminated concrete, the process usually follows four steps:

1. $Fe + 2Cl \rightarrow FeCl_2 + 2e^-$
2. $2e^- + \frac{1}{2}O_2 + H_2O \rightarrow 2OH^-$
3. $FeCl_2 + 2OH \rightarrow Fe(OH)_2 + 2Cl$
4. $2Fe(OH)_2 + \frac{1}{2}O_2 \rightarrow Fe_2O_3 + 2H_2O$

Step 1. Iron in the reinforcing steel and chloride ions in the contaminated concrete combine to form ferrous chloride and free electrons in the reinforcing steel. This steel is acting as the anode.

Step 2. The outer mat of steel is typically connected to the lower mat by ties and spacer, which serve as an electrical connection. The lower mat acts as the cathode where the current leaves the cell and free electrons combine with oxygen and water in the concrete to form hydroxide ions.

Step 3. Hydroxide ions migrate through the concrete back to the anode and combine with ferrous chloride to form ferrous hydroxide and chloride ions.

Step 4. The ferrous hydroxide in anode combines with oxygen to form red iron oxide (rust) and water.

When patching concrete, a section of contaminated concrete is replaced with chloride-free concrete that is usually of a higher pH than the concrete in place. Reinforcing steel that passes through both sections is passing through sections with significantly different corrosion potentials. With this, electrons flow within the bar from the contaminated side into the new patch material. From there the migration of hydroxide ions to the contaminated concrete is very short. As a result, the process of corrosion in the reinforcing steel immediately adjacent to the patch can be accelerated resulting in repairs being required within 2–5 years [2].

Prior to patching concrete the primary mitigation for corrosion is keeping concrete dry by sealing cracks to slow the corrosion process and the migration of any existing chloride ions to the reinforcing steel (Figure 5.5). When patching, using a low-permeability concrete with admixtures to inhibit corrosion will serve to disrupt the action of concrete as an electrolyte. If the reinforcing steel in a patched area is epoxy coated, restoring the integrity of the coating in the repaired area will serve to isolate the steel electrically and protect it from exposure to water and chlorides.

If stronger measures are needed or desired to prevent corrosion around a patched area, cathodic protection may be used. Cathodic protection methods work by reserving the flow of electrons, changing the compromised reinforcing steel into a cathode and protecting it from corrosion. The electrons may be supplied by either a low-voltage direct current supplied to the reinforcing steel through the concrete, an impressed current system; or by a sacrificial material, a galvanic anode.

Galvanic anodes use a dissimilar metal, typically zinc, applied either to the surface of the concrete, or tied to the reinforcing steel directly (Figure 5.6). Since the anodes are sacrificial, they have a limited life; but

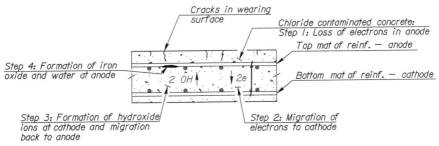

Figure 5.5 Corrosion in Reinforced Concrete.

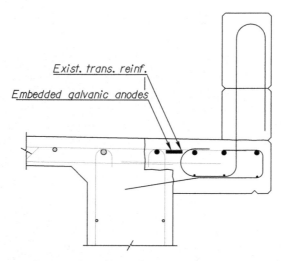

EMBEDDED GALVANIC ANODES: Install anodes on existing transverse reinforcing in the top of slab that are incorporated into the new overhang. Original plans show the longitudinal spacing for this reinforcing is 5 in. Beginning approximately 2 ft. 0 in. from the end of slab, install anodes at 2 ft. 1 in. intervals or every fifth bar.

Figure 5.6 Galvanic Anode.

they are cheaper than the impressed current systems and require little to no maintenance.

5.3.1.6 Drilled and Grouted Reinforcing

If the existing reinforcing steel in a damaged area of reinforced concrete has lost significant section due to corrosion, or if the reinforcing steel was insufficient as originally provided, additional reinforcing steel may be drilled and grouted into the adjacent portion of the concrete member. The technique consists of drilling a hole slightly larger than the new reinforcing bar into the existing concrete and grouting it into place with an epoxy adhesive. The depth of the hole required is typically less than the development length usually used in detailing reinforcing steel. There are two reasons for this, first, the bond strength of the epoxy to the reinforcing and to the concrete is greater than the bond strength between a typical deformed steel reinforcing bar and concrete, and second, the stresses to the concrete are spread across the perimeter of the drilled hole rather than the smaller perimeter of the bar. Information on the depth of hole required to develop the capacity of the bar is typically provided in the manufacturer's literature. Many agencies will also have minimum depths for drilling and grouting in their specifications.

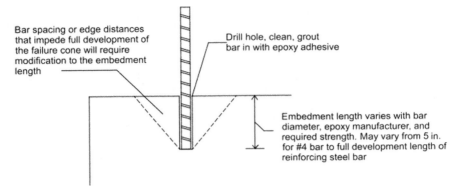

Bar spacing or edge distances that impede full development of the failure cone will require modification to the embedment length

Drill hole, clean, grout bar in with epoxy adhesive

Embedment length varies with bar diameter, epoxy manufacturer, and required strength. May vary from 5 in. for #4 bar to full development length of reinforcing steel bar

Figure 5.7 *Drill and Grout Detail.*

Commonly, drilling and grouting are used to provide reinforcing near the edge of a concrete member, such as for rail post on a deck slab (Figure 5.7). Anchorage of the reinforcing is provided by distribution of stresses along a cone of material surrounding the hole. Where this cone is intersected by a free edge, or an overlapping cone from the anchorage of another bolt, the anchorage is compromised. To compensate reduction factors will need to be applied, which will result in deep holes. Often a thinner concrete member, such as a bridge deck, will lack the necessary depth to fully develop bars of #6 size or greater.

Another consideration for the designer with drilled and grouted reinforcing is the tendency of epoxy to creep under sustained loading. In July 2006, a section of the ceiling of the Interstate 90 connector tunnel in Boston, Massachusetts collapsed onto traffic resulting in fatalities. The cause was the creep of epoxy adhesive in the drilled and grouted anchorage of the ceiling panels. Subsequently, many states prohibited their use where the bars would be under sustained tensile loads.

5.3.1.7 Construction Phasing and Distribution of Stresses

When a section of concrete is removed, the stresses originally borne in that section are redistributed across the member and to adjacent members. The member assumed a dead load deflected shape based on the reduced cross-section. When concrete is replaced in the sections, the dead load stresses in the member are not redistributed into the new concrete until shed by the adjacent sections. That can only occur when those sections are overloaded or if dead load is relieved across the member and reapplied after the new concrete is in place.

When designing repairs, designers must be cognizant only not of limiting the area of concrete removed and length of reinforcing debonded due to capacity requirements, but to the distribution of stresses post repair.

5.3.2 Steel

5.3.2.1 General Discussion

Unlike concrete, steel is a ductile material. It is capable of significant elastic and plastic deformation before failing. Whereas damage to concrete generally manifests itself in the progression of cracking, delaminating, spalling, and eventually fracturing through, damage to steel takes forms that are generally specific to the cause of damage. Cyclic loading will lead to the growth of fatigue cracks. Corrosion will result in the development of corrosion products (i.e., rust) and a loss of section. Impacts to a steel member may lead to excessive plastic deformations at locations where the member is free to deflect and to fractures at locations where the member is restrained. As a consequence, methods of repair on steel members are dependent on the cause of damage.

5.3.2.2 Fatigue Repairs

Other than light corrosion, the most common damage found on steel bridge members is cracking. The most common cause of cracking is fatigue [3]. As discussed in Chapter 2, fatigue cracking occurs in areas of discontinuity or constraint and is driven by cycles of tensile stress. Small cracks may also be found in steel members in and adjacent to welds in areas under compression. Cracks form here from tension resulting from the residual stresses inherent in welding; however, they are rarely of concern since crack growth will stop when the crack tip extends outside of the region of residual stresses.

Upon finding fatigue cracks in a steel member, the first step in addressing them is to determine whether they should be treated, or simply monitored for growth. Any treatment involves expense, and the more significant interventions involve modifications that may affect the load-carrying capacity of the member. If a crack is in a compression area, or is in an area of low-tensile stress and appears to be stable over multiple inspection cycles, the appropriate course of action may simply be continued monitoring.

The goal of treating fatigue cracks is to prevent their further growth. Cracks typically propagate relatively slowly with cycles of loading, but once they reach a critical size they may suddenly grow through to fracture the section. This growth occurs after the stress-intensity factor (K_I) exceeds the critical stress–intensity factor (K_{Ic}) for the material. The stress–intensity factor is a parameter that describes the intensity of the stress field ahead of the crack tip. It is a function of the geometry of the crack, the crack size, and the applied stress. Treatments to arrest crack growth typically address the tensile stresses found at the crack site.

Cracks which have not extended through the thickness of the underlying plate or element may be addressed by surface treatments. Surface cracks will typically be located in, or adjacent to, welds. Cracks in base metal, due to reasons such as out of plane distortion, will manifest as through thickness cracks. Surface treatments address superficial discontinuities and residual tensile stresses. They include: grinding, air peening, ultrasonic impact treatment, and remelting welds.

Grinding involves removing a small portion of the section of a member with an abrasive wheel or bit (Figure 5.8). The limits of the removal include the entirety of the surface crack and are smoothed to avoid discontinuities in the surface profile. The FHWA *Manual of Repair and Retrofit of Fatigue Cracks in Steel Bridges* recommends that areas gouged out should be tapered at a minimum of 2.5:1, that is, a gouge 0.5 in. deep should be tapered to the surface

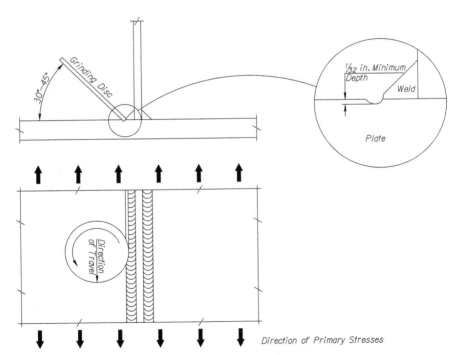

Figure 5.8 *Grinding. (Adapted from FHWA Manual [4]).*

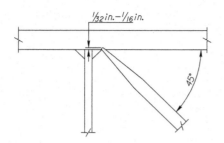

PROCEDURE:

1. Adjust air pressure regulator to 40 psi.

2. Attach air hammer with proper peening tip to the air line.

3. Place tip of peening tool on weld toe and press trigger on air hammer. Keep moving the hammer along the weld toe at a rate of 3 in./min. Make 6 passes along the weld toe. Depth of indentation should be approximately $^1\!/_{32}$ in.

4. Lightly grind the peened surface to remove any lap marks.

Figure 5.9 *Air Peening. (Adapted from FHWA Manual [4]).*

over a length of 2.5 in. The edges of the gouge should be smoothed and rounded. The finish grinder should move parallel to the direction of primary stresses so that grind marks are also parallel to the direction of primary stress.

Air peening involves impacting the treated area with a pneumatic hammer utilizing a blunt tip to induce compressive stresses at the site (Figure 5.9). According to studies cited by the FHWA *Manual of Repair and Retrofit of Fatigue Cracks in Steel Bridges* peening can be used to repair cracks up to 0.12 in. (3 mm) deep. Making six passes using a hammer with 40 psi air pressure will consistently induce a depth of deformed grains of over 0.02 in. (0.5 mm). This induces a depth of compressive stresses 0.04–0.08 in. (1.0–2.0 mm) deep. Treated areas can be considered to improve one fatigue detail category in those used by the AASHTO LRFD Bridge Design Specifications.

Ultrasonic impact treatment (UIT) is a proprietary method of inducing compressive stresses on the surface of a material in a manner similar to that of air peening, but with smaller, lighter, and quieter equipment. Impacts are provided by low amplitude, high-frequency (27 kHz) displacements. UIT was developed in 1972 by Dr Efim Statnikov for use on Soviet submarines and demonstrated to the Federal Highway Administration in 1996 at the Turner–Fairbank Research Center [5]. It has been used on several highway bridges in the United States since that time.

A length of cracked weld may be removed and replaced, as will be discussed later in the chapter. A less invasive treatment for surface cracking at welds is to remelt the weld with a gas tungsten arc (alternately known as the tungsten inert gas process). The FHWA *Manual of Repair and Retrofit of Fatigue Cracks in Steel Bridges* reports that this process may be used to repair cracks up to 3/16 in. (5 mm) deep. Similar to welding, this process requires a qualified and certified operator to perform.

A length of cracked weld may be removed by gouging out with the air–arc method and then replaced by a new weld. However, this requires an analysis to ensure that the structure is stable during the removal of the damaged weld and surface treatment, preheating and postheating as per the AWS D1.5 guidelines. A radiograph of the repair is typically required after completion. Due to the expense of this operation, it is rarely done if another treatment is applicable.

Through thickness fatigue cracks are most often addressed by drilling holes at the ends of the crack. The ubiquity of its use is due, at least in part, to the relative ease of implementation. The only equipment, outside of the typical equipment involved in the safety inspection of steel bridges, required is a portable drill with an electromagnetic base (mag drill).

Removing the end of the crack removes the sharp discontinuity and replaces it with a smooth edge that is more conducive to the flow of stress. Care must be taken to capture the crack tip and the adjacent plastic zone ahead of the tip. This requires a minimum hole diameter of 1 in. and perhaps up to 4 in. Large hole diameters must be evaluated for their effect on the structural capacity of the member (Figure 5.10).

FHWA reports [3] that the minimum diameter hole needed at the tip of a crack with is contained completely within a plate may be calculated as:

$$D = \frac{S_r \pi a}{8\sigma_y} \geq 1.0 \text{ in.}$$

where, D, hole diameter in in.; S_r, stress range in ksi; a, one half of crack size; and σ_y, yield stress of material in ksi.

Note, for a crack at a free edge, the calculated minimum hole diameter is increased by 25%.

The edges of the hole should be dressed with a burr grinder to remove rough edges and should be protected with a paint system. Compressive stresses may be induced at the location by either placing and tensioning a

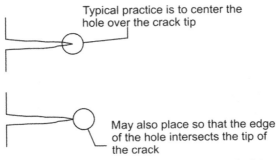

Figure 5.10 *Location of Crack Tip Hole.* Note: The end of the crack should be determined by mag-particle inspection rather than visually, when possible.

high-strength (A325 or A490) bolt and washers, or driving a drift pin into the hole with a blow sufficient to deform the lip of the hole.

A through thickness crack may also be treated by splicing over the crack with additional material. Plates, or rolled shapes such as angles, are typically added symmetrically to both sides of the cracked element to avoid introducing any eccentricity. The design of the repair differs slightly from that of a bolted splice in new construction. Since the cracking is driven by tensile stresses, the added material serves to replace the tensile capacity of the cracked element. For a crack through a web plate, plates at least as thick as half the web thickness with sufficient bolts to match the tensile capacity of the plate should be sufficient. It is not recommended to attach plates for the repair by welding unless there is no room to develop a bolted connection. The repair should be detailed as a slip critical connection, which will require a degree of surface preparation for the faying surfaces.

The section on fatigue evaluation in Chapter 3 notes that fatigue damage may be categorized as either load-induced or distortion-induced. Load-induced fatigue cracks are driven by primary stresses in a structural member, for example the cracking that may be found at the ends of welded cover plates on the tension flanges of bridge girders (Figure 5.11). Distortion-induced fatigue is caused by the out-of-plane motion of secondary framing attached to primary members, that is, a cross-frame attached to a girder.

Distortion-induced fatigue damage most commonly occurs in the web plates of girders at the gap between the flange and the attachment of a connection stiffener to the web (Figure 5.12). As truck traffic on adjacent

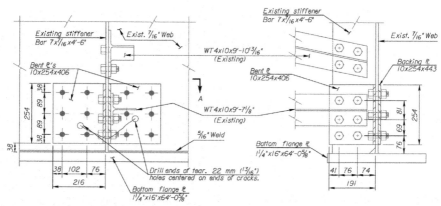

Figure 5.11 *Bolted Splice Over Fatigue Crack.*

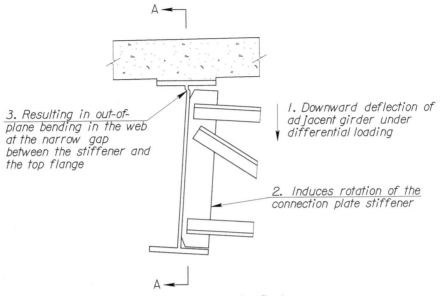

3. Resulting in out-of-plane bending in the web at the narrow gap between the stiffener and the top flange

1. Downward deflection of adjacent girder under differential loading

2. Induces rotation of the connection plate stiffener

Figure 5.12 *Web Gap Distress From Differential Deflection.*

girders causes them to deflect downward, the secondary framing is engaged, resulting in a tensile force being applied at one end of the connection stiffener. Prior to 1985, the design code did not require this stiffener to be attached to the tension flange of girders. The justification for not attaching the stiffener was to avoid having a transverse weld in a tension area. When not attached, the tensile force must be resisted solely by the web. This commonly results in horseshoe shaped crack around the end of the connection stiffener attachment to the web (Figure 5.13).

The first response to these cracks is typically to drill holes at the ends. The FHWA *Manual of Repair and Retrofit of Fatigue Cracks in Steel Bridges* reports that research at Lehigh University found that hole drilling was sufficient only if the in-plane bending stress at the site was less than 6 ksi and the out-of-plane stress was less than 15 ksi. Further, hole drilling was found to be insufficient to stop crack propagation at lesser stresses on skewed bridges. For bridges where hole drilling is insufficient, the repair method for web gap cracking follows one of the three approaches:

1. Soften the web gap by removing material;
2. Stiffen the web gap by positive attachment to the flange; and
3. Distributing the out-of-plane force over a wider area of the web with a backing plate.

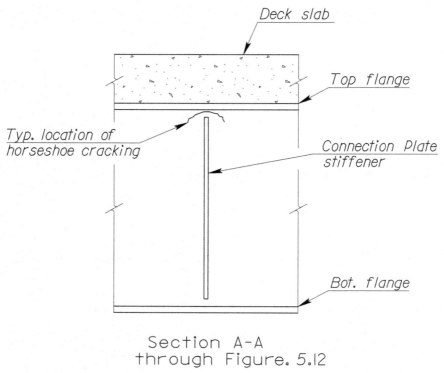

Section A-A
through Figure. 5.12

Figure 5.13 *Horseshoe Crack at Web Gap.*

Softening of the web may be done by removing a portion of the connection plate to increase the web gap length (Figure 5.14). It has been done with mixed results at locations where the secondary framing is cross-frames and diaphragms. It is the repair recommended by the FHWA *Manual of Repair and Retrofit of Fatigue Cracks in Steel Bridges* for the connection of deep floor beams to girders. It is recommended that the gap be a minimum of one-sixth of the web depth, not to exceed 15 in.

Softening may also be done by drilling large (>3 in.) holes in the web adjacent to the connection stiffener. The capacity of the member should be reviewed prior to specifying this repair, as a hole of this size may reduce the member's capacity (Figure 5.15).

A more common approach to addressing web gap cracking is to stiffen the gap by positively attaching the connection plate to the flange, as required in current design practice. Before applying this repair, the live load stress range for Load and Resistance Factor Design (LRFD) fatigue load cases at the specific location should be reviewed to determine if they are acceptable for the fatigue detail category for the repair.

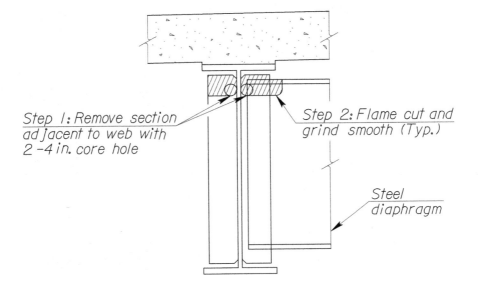

Step 1: Remove section adjacent to web with 2 – 4 in. core hole

Step 2: Flame cut and grind smooth (Typ.)

Steel diaphragm

Figure. 5.14

Figure 5.14 Cutting Stiffener to Soften Web Gap.

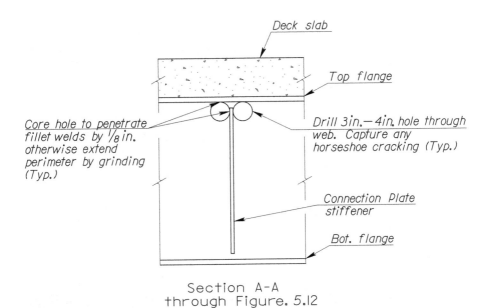

Deck slab

Top flange

Core hole to penetrate fillet welds by 1/8 in. otherwise extend perimeter by grinding (Typ.)

Drill 3in.– 4in. hole through web. Capture any horseshoe cracking (Typ.)

Connection Plate stiffener

Bot. flange

Section A-A through Figure. 5.12

Figure 5.15 Large Hole Retrofit for Web Gap Softening.

When conducting the repair, the traffic lane above the girder should be closed, to prevent vibrations during welding. If welding the connection stiffener to the flange, care must be taken that all surface preparation measures and pre- and postheating protocols for welding per AWS D1.5 are followed. The weld itself should be sized to resist a force of 20 kips per AASHTO LRFD Bridge Design Specification 6.6.1.3.1 unless an analysis indicates otherwise. Note that the load effects in the cross-frames of highly skewed or curved bridges should be investigated in particular.

If the live load stress ranges at the location do not permit a welded connection, or if the preference of the owner is simply to field bolt whenever possible, then the connection plate may be attached with either angles or a WT section bolted to both the stiffener and the flange. If the top face of the top flange is accessible (such as when the deck is removed for a redecking project), high-strength bolts through the flange may be used. If the top flange is only accessible from the bottom face, then bolts may be placed by:

- Drilling and tapping holes into the flange. FHWA recommends a minimum embedment of 1 in.
- Drilling through the flange and grouting into the concrete deck above. Embedment into the deck is as required to develop the threaded rod.
- Attaching a threaded rod with a stud-welding gun. Note, the fatigue detail category of the weld is C.

Commonly, a hybrid attachment is used that is bolted to the flange and welded to the stiffener. Bolting to the flange prevents any cracking that might develop in the weld to the stiffener from propagating into the girder (Figures 5.16 and 5.17).

Researchers at the University of Kansas have developed another style of repair where the cross-frame attachment to the connection stiffener is reinforced by bolting on angles, which are bolted to a backing plate on the opposite side of the web [6]. This allows distortion to be resisted by the section of web engaged by the backing plate, rather than a single point of attachment. The stress ranges in the web are reduced enough to prevent crack growth (Figure 5.18).

5.3.2.3 Corrosion Repairs

Corrosion of steel results in a loss of section in the member. Since the specific volume of the corrosion products of iron vary from two (for Fe_2O_3) to five (for $Fe_2O_3 3H_2O$) times that of iron, rust has a significantly larger volume that the original steel. As structural steel corrodes the material will expand and will flake away. This damaged surface may further serve to trap

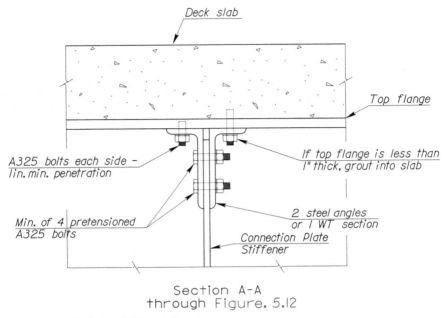

Figure 5.16 *Web Gap Stiffening Repair.*

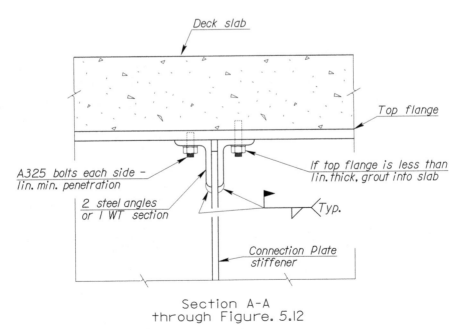

Figure 5.17 *Hybrid Web Gap Stiffening Repair.*

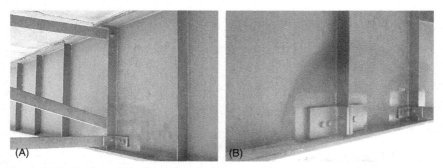

Figure 5.18 *Angle and Backing Plate Repair.*

moisture resulting in more corrosion. If the process continues, the section loss may become significant.

Bridge inspectors need not only to ascertain the location and amount of any section loss, but should inspect surrounding area for additional damage that the loss in member capacity due to corrosion might have caused. The shedding of load may have increased in stresses in surrounding areas sufficiently to compromise welds, plates, or other member components. It is important to check for indications, such as buckling or plastic deformations, that a member might have been overloaded. Plates or member components that plastically deformed due to overloading caused by loss of section are dealt with in the same manner as those overloaded as a result of impact loads. The treatment of plastic deformation is discussed in the Section 5.3.2.4.

The initial effort to repair the effects of corrosion is to stop its growth, and therefore prevent any further loss of capacity. This entails removing existing corrosion products and protecting the area with a paint system, and addressing any contributing deficiencies in immediately environment, such as leaking expansion joints. Painting may be done as part of repainting the entire superstructure or by spot painting, as discussed in Chapter 4. If the loss of section is not great enough to significantly impact the capacity of the member, this may be the limit of the scope of the repair (Figure 5.19).

If the load-carrying capacity has been compromised and must be restored, the repair entails replacing the lost area of steel. Depending on the amount of section loss and the cost in terms of material and effort, either the section may be augmented with new steel or the member may be replaced. Replacement of the member itself may be in whole or in part. If member replacement is the economical choice, the limits of the replacement and

Figure 5.19 *Section Loss Inside of Truss Chord.*

providing for possible shoring requirements for the structure must be considered.

When replacing the lost steel, the designer must not only consider the area required to transmit load, but also the effect of the new piece on the stiffness of the member. A new plate with a yield stress of 50 ksi may safely carry the same amount of load in tension as an existing plate of a yield stress of 36 ksi with only 72% (=36/50) of the area. However, using a smaller plate in replacing an existing tension flange in a steel beam will result in an increase in live load deflections because the reduction in section will result in a reduction in the moment of inertia (i.e., the stiffness) of the section.

Designing the connection of a new section of steel to the existing requires careful consideration. Often constraints in the geometry at a repair location limit the space available to develop a connection. A welded connection may achieve the required capacity in a smaller area than will a bolted connection; however, welding will induce residual stresses into the area. This may induce fatigue cracking if the location is restrained from deflection or rotation. Additionally, welding may be significantly more costly do to than bolting since a certified welder and subsequent material inspection of the welding is required. For these reasons, in maintenance work, the preference is to use bolted connections if it is possible to develop the required capacity at the location of the repair (Figure 5.20).

When designing a bolted splice connection, consideration must be given to the degree of surface preparation possible at the site of the repair. The default bolted connection type for the design of a modern steel highway bridge is slip critical. In a slip critical connection, adjoining plates are pulled into contact with sufficient force by the tension in high strength bolts to develop sufficient friction that the plates will not slip from each other when the member deflects. The alternative to a slip critical connection is a bearing connection. In a bearing connection the adjoining plates slip relative to each other when the member deflects until the bolts contact the edges of the bolt hole. A slip critical connection precludes fatiguing of the bolts or unwanted rotation between plates under normal service loads. However, to develop the necessary friction, the contacting surfaces must meet a certain surface profile. If it is not possible to achieve the required surface preparation, a bearing connection must be used.

Figure 5.20 *Repair Splice in Truss Chord.*

5.3.2.4 Impact Repairs

While damage from fatigue and corrosion generally starts as a small defect, which grows over time, the damage from vehicular impacts occurs in an instant. The bent and twisted girders make for dramatic images. Impact loads typically result in two forms of damage to steel members: cracking and plastic deformation. Under a very severe impact, the most severe cracks may be complete fractures of elements or components of elements. More often, cracks will appear at welds or at points of restraint (such as cross-frame locations), and will be dealt with in the same manner as fatigue-induced cracks. Plastic deformation is the type of damage that is characteristic of impact for bridges in service.

Plastic deformation occurs as steel is stressed beyond its elastic limit, that is, past yield stress. Once the external load driving the stress is removed, a steel member will rebound for the elastic portion of the deformation, but the deformation that was incurred past yielding will remain permanently, unless corrected by a maintenance action. Structural members that are twisted or swept out of their intended shape no longer have their original load-carrying capacity. In the FHWA guide *Heat Straightening Repairs of Damaged Steel Bridges* [4], the authors calculated the reduction in the effective section modulus for each angle of rotation of the web and bottom flange about a center at the top of the web for W24 × 76 and W10 × 39 beams. For these rolled sections, a 10° rotation resulted in a 10% reduction in effective section modulus, and therefore in moment capacity. They offered that a 10% reduction in effective section modulus was a good general guideline as to when straightening of the member would be required for it to remain in place.

Plastic deformation by mechanical means is often used in manufacturing processes to improve the qualities of steel for certain uses. Cold working steel in the shop results in increasing the yield strength, the ultimate tensile strength and the hardness of the steel. However, this does come at the expense of ductility; and, due to the Bauschinger effect, increases in tensile yield strength may be offset by reductions of compressive yield strength. Plastic deformation in the fabrication shop is controlled; vehicular impact loadings are applied suddenly and violently. Residual stresses are induced in steel members that may exacerbate crack formation and growth at flaws introduced by the impact and at previously existing unknown flaws in the member. In fabrication shops, a maximum strain value of 5% is traditionally used as a limit for cold working before heat treatment is required to alleviate the effects of residual stress [7].

After any impact, the bridge must be inspected and any cracks found should be dealt with as per the techniques discussed in the previous section. Any notches, gouges, or sharp defects which may provide a flaw for the initiation of future crack propagation should be ground out and dressed if retaining the member, regardless of whether any straightening is done and prior to any straightening operations (Figure 5.21). For the repair of gross plastic deformations there are four options:

1. Leave the gross deformation of the members as they are.
2. Mechanically straighten the members.
3. Heat-straighten the members.
4. Replace the members.

The decision whether to leave a deformed member as is, is dependent on an evaluation of the structural capacity of the member. Although, a limit of a 10% reduction in effective section modulus provides a general guideline, there are other considerations. For a structure with a low load rating, a 10% reduction in moment capacity of a member may not be sustainable. If the member is a column or a pile in a pile bent, there are $P - \Delta$ effects to consider.

Whether a member may be straightened by mechanical means depends on if it is a primary load–carrying member and the amount of strain incurred by the member. National Cooperative Highway Research Program (NCHRP) Report 271 provides guidelines for the applicability of mechanical straightening[1] [8]. It allows mechanical straightening regardless of the

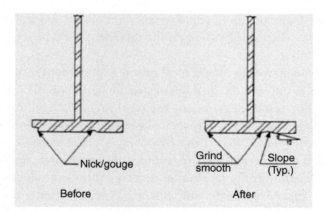

Figure 5.21 *Grinding Out Small Nicks and Gouges. (Adapted from Federal Highway Administration. Bridge maintenance training. Washington, DC; US Department of Transportation; 2003).*

[1] The guidelines in NCHRP Report 271 were also intended to apply to heat-straightening repairs. However, they have been superseded by subsequent guidelines specific to heat straightening by FHWA and later NCHRP research.

amount of strain in the member for compression members and for secondary members. Primary tension members may be straightened if the nominal strain from damage is equal to or less than 5%. If the damage occurs in a location on the member with fatigue details that are less than C, then addition material must be added to strengthen the member after straightening if the strain if equal to or greater than 15 times the yield strain of the steel. For A36 steel this is a strain of approximately 1.5%. The amount of added material is to be at least 50% of the material in place.

For primary tension members with nominal strains from damage greater than 5%, mechanical straightening is allowed if the member is strengthened with material that would provide an additional 100% of the existing capacity. Although NCHRP Report 271 does not apply a limit to straightening compression members, cold mechanical straightening does not alleviate the effects of residual stresses. These stresses may negatively affect the buckling capacity of a member, therefore if mechanical straightening is used, it would be best to at least apply the limits for tension members to all primary bridge members.

The nominal strain for deformation of member can be estimated from direct measurement of the damage. NCHRP Report 271 provides the derivation relating the offset measured at the point of damage from the original alignment of member to radius of curvature of the damaged shape, R:

$$\frac{1}{R} = \frac{y_{r-1} - y_r + y_{r+1}}{L^2}$$

Where L is an arbitrary distance chosen to be within the yielded zone. The strain, ε, is related to R by:

$$\varepsilon = \frac{y_{max}}{R}$$

Where y_{max} is the distance from the centroid of the element in bending to the extreme fiber of the element. In this case, it is half of the thickness (or width, depending on the orientation of the damage) of the piece bent (Figure 5.22).

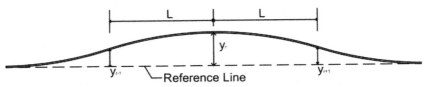

Figure 5.22 Offset Measurements of Calculate Radius of Curvature. (Adapted from NCHRP Report 604 [9]).

Mechanical straightening may be done with the addition of heat to reduce the yield strength of the member at the time of the application of force. Guidelines subsequent to NCHRP Report 271 strongly recommend against using hot mechanical straightening in the field due to possible negative effects of uncontrolled heating on the material properties of steel.

The recommended method of straightening plastically deformed steel members is heat straightening. This process uses patterns of heating in cycles to produce movement of the steel member. Heating steel reduces its yield strength during the time the heat is applied. For modern bridge steels this begins around 700°F. The heated section of steel experiences plastic flow and thermal expansion while the surrounding cooler steel acts to constrain the thermal expansion. The result is a slight thickening, or upsetting, of the heated section. When heat is removed and the section cools, the thermal contraction produces movement.

The reader is referred to the FHWA guide *Heat Straightening Repairs of Damaged Steel Bridges* for discussion of particular heating patterns as they related to correcting deformation. Because of the unique geometry of every impact and the unique pattern of residual stresses in the affected steel members, the application of heat straightening is dependent on the skill and the experience of the operator. As long as steel is not heated past its lower critical (phase transition) temperature, approximately 1300°F for bridge steels, and is allowed to air cool (no quenching), its material properties in the final repaired state are not profoundly diminished (Figure 5.23).

The process of straightening a member may be accelerated if it is restrained from the undesired movement associated with expansion at the beginning of the heating cycle. Note that the intent of the application of

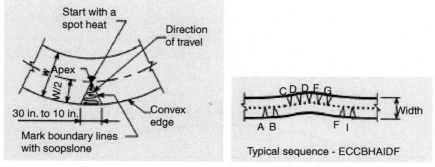

Figure 5.23 *Typical Vee Heat Pattern.* *(Adapted from Federal Highway Administration. Bridge maintenance training. Washington, DC; US Department of Transportation; 2003).*

Figure 5.24 *Restraint Applied to Girder Being Heat Straightened.*

force during heat straightening is only to restrain, not to cause movement (Figure 5.24). To that end, the FHWA guide provides procedures for calculating the maximum allowable force. The moments produced by restraint jacks on the member should be less than half of the plastic moment capacity of the section less any residual moments from the impact as determined by a plastic analysis. In lieu of calculating residual moments, a limit of one-quarter of the plastic moment capacity of the section may be used.

If an accurate modeling of the deformed structure is not available, or if the restraining jacking force limit must be estimated in the field, the guide provides that the force may be limited to that which produces a deflection that produces maximum stress equal to 50% of yield. For continuous beams this may be estimated as a maximum deflection of:

$$\delta_{max} = \frac{1}{y_{max}}\left(\frac{L}{200}\right)^2 \quad \text{For Grade 36 steel}$$

$$\delta_{max} = \frac{1}{y_{max}}\left(\frac{L}{170}\right)^2 \quad \text{For Grade 50 steel}$$

Where y_{max} is the distance from the centroid of the element in bending to the extreme fiber of the element and L is the distance between diaphragms, cross-frames, or other lateral support.

Research on heat straightening subsequent to publication of the FHWA guide is found in NCHRP Report 604. It is reported that heat straightening has effects on the material properties of steel mostly similar to that of mechanical straightening, increasing the yield strength and lowering the ductility of the steel. Also, some residual stresses are introduced into the steel by the heat-straightening process. However, heat straightening is applicable for repairs of much larger strains than repairable by mechanical means. NCHRP Report 604 recommends the procedure for strains 150 times the yield strain of the steel, and reports that it has been successfully applied for strains slightly over 200 times yield strain.

5.3.2.5 Fire Damage

Even if a vehicular crash at a bridge site does not result in a direct impact to steel members, the members may be damaged by subsequent fire. As part of NCHRP Project No. 12-85, researchers from Virginia Polytechnic evaluated the hazard posed to highway bridges by fire in the United States [10]. Reviewing crash data from databases on vehicular crashes maintained by the National Highway Traffic Safety Administration they found an average of 105 crashes each year involving bridges and fires. A review of fire statistics maintained by the Federal Emergency Management Agency, found that there are an average of 243 vehicle fires on bridges each year. Compared to the total number of bridges in the United States (approximately 600,000) the estimates of fires per year on bridges are in fairly good agreement.

Their review of a database of bridge collapses in the United States since 1960, maintained by the New York State Department of Transportation, found that out of 1746 collapses, 50 were caused by fire. Although fire as a cause of collapse with small relative to hydraulic events (1006), it was similar in number to collapses contributed to corrosion (64) and more than that due to earthquakes or other natural events.

As previously discussed in the review of heat straightening, the yield strength of steel decreases at high temperatures. Comparing values from test data and models of steel yield strength as a function of temperature from the American Society of Civil Engineers and from the Eurocode for Design of Steel Structures, the Virginia Tech researchers show that the yield strength of steel at 1100°F is 40–50% of what it is at 70°F. Fires fueled by petroleum may readily reach 1800°F within a few minutes. The fire modeled in ASTM

International Standard E1529, *Standard Test Methods for Determining Effects of Large Hydrocarbon Pool Fires on Structural Members and Assemblies*, climbs to 2000°F within 5 min (Figure 5.25).

The primary damage to steel members from fire is plastic deformation. However, heat straightening is typically not an option to repair members deformed from fire. At temperatures over 1300°F, steel undergoes a phase transition and potential changes in material properties. Even if the temperature of the steel remained below this, if water was sprayed on the steel while the fire was being suppressed, the rapid cooling will have a delirious effect on member properties. For substantially deformed members, the only repair option is replacement.

5.4 SUBSTANTIAL MAINTENANCE ACTIONS

The various elements of a bridge are subject to different levels of exposure to the environment and wear by traffic loads. As a consequence, each element will have a different expected service life. Substantial maintenance work is typically programmed for specific bridge elements, such work is discussed later in the chapter.

Figure 5.25 *Fire Damaged Steel Girders. (Adapted from Federal Highway Administration. Bridge Inspector's Reference Manual. FHWA-NHI 12-049. Washington, DC; US Department of Transportation; 2012).*

5.4.1 Decks and Railing

The elements of most bridge types that are exposed directly to traffic are the deck and railing. The majority of highway bridge substantial maintenance work is conducted to maintain the wearing surface of the deck. As discussed in Section 2.3.2.2, decks on the vast majority of modern highway bridges are constructed from reinforced concrete. These decks may be constructed from one course of concrete or two courses where the second course forms a nonstructural overlay (typically 1.5–2.5 in. thick). They may also be formed using prestressed concrete panels (typically 4 in. thick). Almost all concrete deck maintenance involves removing and replacing deteriorated concrete. This work is described in the Section 5.3.1.3 on concrete patching and surface repair. When applied to the wearing surface of a concrete deck, the Type B patch is often referred to as partial-depth patching and the Type C patch is often referred to as full-depth patching (Figure 5.26).

Wearing surface patches are poured into place. Partial-depth patches require no formwork and are typically substantially cheaper on a unit price basis than are full-depth patches or other types of concrete surface repair. Concrete deck patching is usually done with the standard concrete mix used by the bridge owner (Figure 5.27). The primary consideration for using a special mix is when a quicker cure time is needed to minimize impact to traffic. Whereas a typical concrete mix is allowed to cure for several days (the exact number varies between different bridge owners) before that portion of deck is opened to traffic, a high early strength mix may achieve sufficient strength in less than a day. However, there is a greater cost for higher early mixes and their durability may not match that of the standard concrete mix.

When scoping a deck-patching project, it is important to determine if the bridge type is either a structural deck or composite girder. For a structural deck, the deck is the primary load-carrying element from pier to pier, while for the composite beam bridge the deck participates in composite

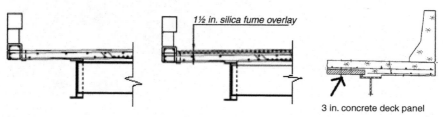

Figure 5.26 *Section of One Course and Two Course Decks and Deck With Prestressed Concrete Panel.*

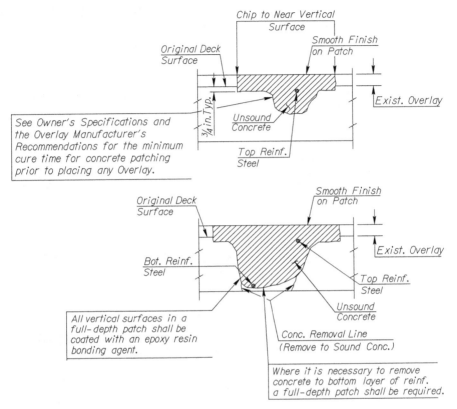

Figure 5.27 *Concrete Deck Patching.*

action with the beam or girder in carrying load from pier to pier. For either of these bridges, the amount of longitudinal (parallel to the direction of traffic) reinforcing in the top mat exposed and debonded at any one time must be limited. On noncomposite beam and girder bridges, the deck acts to span between girders. The primary load-carrying reinforcing is in the bottom mat, so the amount of top mat steel exposed at any one time for partial-depth patching is not as critical (Figures 5.28 and 5.29).

Replacing deteriorated concrete on the underside of the deck or on the face of the railing involves placing on an overhead or on a vertical surface and will be more expensive on a unit cost basis than will deck patching. The concrete may be poured and formed, or may be placed as shotcrete, depending on the means and methods available to the construction contractor and on the surface finish required (Figure 5.30).

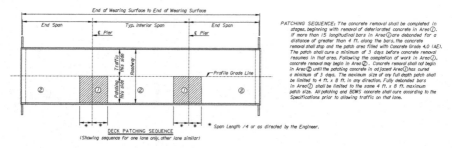

Figure 5.28 *Deck-Patching Sequence on Structural Slab or Composite Girder Bridge.*

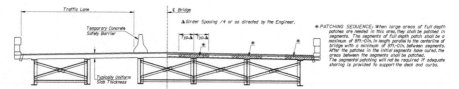

Figure 5.29 *Deck-Patching Sequence on a Girder Bridge.*

For decks that receive a large volume of traffic, or for decks with extensive deterioration of the wearing surface, an overlay may be an appropriate scope of work. Overlays may be rigid, thin bonded, or bituminous. A rigid overlay is a layer of unreinforced concrete applied across the entire area of the deck to form a sacrificial wearing surface. The addition area of concrete is not considered to contribute to the structural capacity of the deck. Rigid overlays may be applied as part of the original construction of a reinforced concrete deck (two-course deck) in locations of high traffic. Rigid overlays applied as a maintenance action may be slightly thinner or thicker (1–3 in.) than those used in new construction as required to meet existing pavement elevation and slope constraints.

Other than durability, the key characteristic desired of material used in a rigid overlay is low permeability. To that end, the typical Portland cement concrete mix is often modified with either latex or a supplemental cementitious material, such as silica fume, to reduce permeability. Similar to deck patching materials, admixtures may also be used to accelerate the curing of the overlay material to minimize traffic impacts.

For sections of the deck that require Type A or B patching, the material for the rigid overlay may serve as the patching material. This may serve to expedite the repair work.

Thin-bonded overlays are multilayer overlays of polymer concrete material (typically either epoxy or polyester) that serve to seal the wearing surface from water intrusion while provide a suitable surface friction for traffic

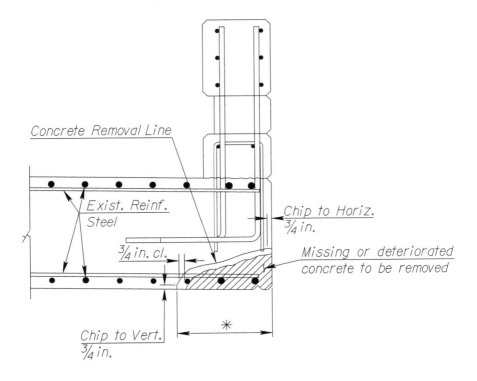

Concrete Removal Line

Exist. Reinf.
Steel

Chip to Horiz.
¾ in.

¾ in. cl.

Missing or deteriorated
concrete to be removed

Chip to Vert.
¾ in.

✳

✳ Pay area for Concrete Surface Repair is
this width times the length of the repair.

Figure 5.30 *Edge of Slab/Overhang Repair.*

operation. The polymer material seeps into the prepared substrate of the concrete wearing surface. There is a layer of fine aggregate broadcast across the top to provide friction. The thickness of the overlay and aggregate is typically around 3/8 in.; however, the polyester polymer concrete may also be used in thicker lifts as a rigid overlay material. It is more expensive than the typical Portland cement concrete overlay materials and so is not used as much. It may be used when increased durability is required.

A thin-bonded overlay may also be used as a sealer to salvage an existing rigid overlay that has begun to crack and debond. If using it for this reason, it is essential to apply it before substantial deterioration of the existing rigid overlay.

Bituminous overlays on reinforced concrete decks are more popular in Canada and in parts of Europe than in the United States (Figure 5.31). Bituminous materials are porous and, if a waterproofing membrane is not in place between the overlay material and the reinforced concrete, then drainage

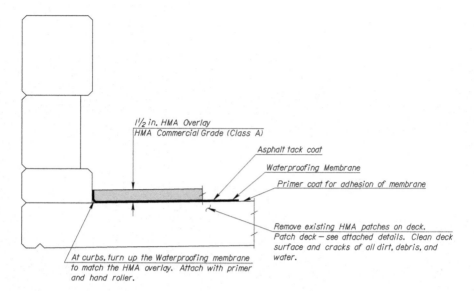

Figure 5.31 *Bituminous Overlay on Bridge Deck.*

from the wearing surface will pass through the overlay and sit atop the concrete. A bituminous overlay may be applied to a deck at the end of its expected life, when subsequent deck deterioration is less of a concern, to facilitate rideability.

The correct application of a waterproofing membrane to the deck can mitigate damage to the deck from drainage and there are several examples of good long-term performance from such systems. However, any holiday (i.e., tear) in membrane provide an opportunity for deck deterioration, which may be exacerbated by the trapping of any water that passes through holiday by the surrounding intact membrane. This concern, along with concerns about the inspectability of the deck when covered with bituminous material, has impeded wide acceptance of bituminous bridge deck overlays in the United States. Where used, they are used mainly on lower volume bridges.

The other types of deck construction on highway bridges are steel grid, timber, and FRP. Steel grid decks are used primarily on long span structures where the savings in material weight result in less loading on the span structure. The main substantial maintenance issue with these decks is repairing broken grids and steel supports underneath. The steel will fracture due to corrosion and loading from traffic and from thermal expansion and contraction. It should be noted that there was a period in the early 1980s when grid deck panels made from weathering steel were available. Weathering steel is not suited for use in this application, or for expansion joints or any other use where continuous

Figure 5.32 *Repair of Steel Grid Deck.*

exposure to water and deicing salts is expected. Continuous exposure to moisture will drive corrosion in weathering steel past the formation of its characteristic protective patina. For a weathering steel grid deck, the result is not only loss of section to corrosion, but growth of the deck itself. As a result, the deck will bow upward requiring repair by cutting the deck, realigning it and reattaching (Figure 5.32).

Timber decks may be found on low-volume bridges. Typically in park areas or as access to properties adjacent to the highway. Repair consists of replacing deteriorated sections or splicing (Figure 5.33).

Decks formed from extruded sections of FRP are still very rare in the United States and have generally been constructed as tests of the technology. Results to date have been mixed. Damage requiring repair generally consists of delaminations of the material, which must be repaired to the particular manufacturer's specifications.

Aside from the repair of damaged concrete as discussed earlier, the most common substantial maintenance action for bridge rails is replacement of sections damaged by traffic impacts. Damaged sections are typically replaced in kind with the original rail. When scoping such a rail, consideration must

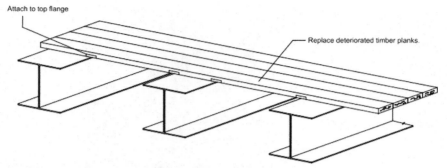

Figure 5.33 Replace Timber Planks on Steel Beam Bridge.

be made of the condition of the underlying edge of the deck. Since deck drainage is carried down the curb line adjacent to rail, this is often the location of the greatest quantity of deterioration of the deck. If deterioration is sufficiently advanced, the edge of the deck may be to be removed and replaced as part of the rail repair (Figure 5.34).

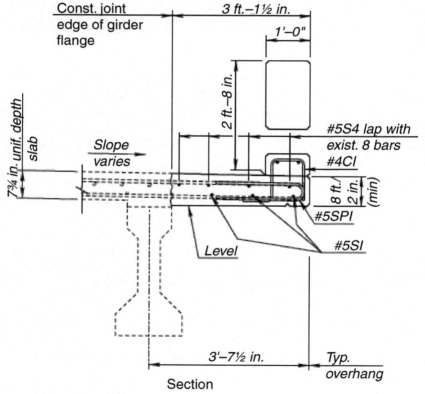

Figure 5.34 Edge of Slab/Rail Replacement.

5.4.2 Expansion Joints

After deck work, the issue most often addressed by substantial maintenance work is the repair and replacement of deck expansion joints. These joints allow for the thermal expansion and contraction of the superstructure independent of the substructure. They are sized to allow for this movement over a range of temperatures specified in the design code. For the Midwestern United States, the AASHTO LRFD Bridge Design Specifications require that a bridge with a steel superstructure accommodate movement resulting from a range of temperatures of −30 to 120°F; for a concrete superstructure the range is 0–80°F. The temperature of a concrete superstructure is presumed to respond more slowly to the ambient temperature than a steel superstructure. While providing for this movement, a joint must also limit the transverse gap across the deck to a maximum opening of 4 in.

Deck expansion joints may be open or sealed. Open joints allow water to drain through. Water may be intercepted and conveyed away by a trough, or may fall through onto the substructure below. The latter design is more common in regions which do not use deicing salts on the highway on a regular basis in the winter. It was also more common in the past than it is currently. It is less popular now because of the deterioration to bearing seats and bearing devices caused by drainage (Figures 5.35 and 5.36).

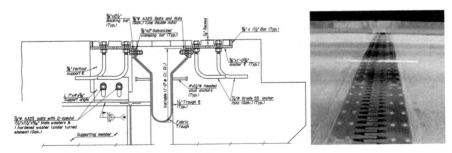

Figure 5.35 *Finger Plate Expansion Joint.*

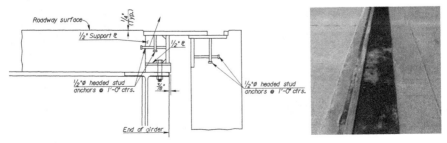

Figure 5.36 *Sliding Plate Expansion Joint.*

Sealed joints use a gland or membrane to prevent water from traveling through. Instead, water is conveyed down the gland, following the cross-slope of the road, to the curb line. If sealed joints are not kept clean of debris, they can tear and allow water onto the substructure (Figures 5.37 and 5.38).

Repair work for expansion device usually has one or more of the following scopes:

• Repair of damaged concrete adjacent to the device;
• Replacement of the gland or membrane for a sealed joint; and
• Adjustment of an assembly joint for rideability.

The concrete adjacent to a joint is subject to impact load from tires traversing the joint and to stresses from restraining the motion of the joint. This is a common area for concrete patchwork. Some joints may be placed with a block out in the adjacent concrete for placement of elastomeric or polyester polymer concrete, either of which should have more impact resistance than standard Portland cement concrete (Figure 5.39).

Joint glands may be replaced in kind, however, often for strip seals and for modular devices, either the gland or personnel with the experience to replace glands are not available. In this case, glands may be replaced with compressible membrane joint seals. Such seals will not be a durable as glands, but are relatively inexpensive and easy to install (Figure 5.40).

During their service life, joints may begin to protrude into traffic due to either differential settlement between the different sides of the device or corrosion of steel under the device pushing it upward. For a steel device such as a finger plate, this may be addressed by grinding the top of the plate to match the vertical alignment of the road grade.

More often, the scope of substantial maintenance work for an expansion device is to replace it. Replacement may be the best way to address

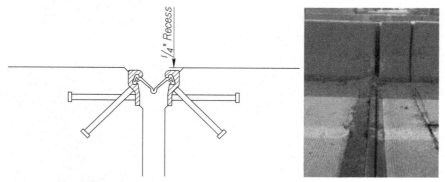

Figure 5.37 *Strip Seal Expansion Joint.*

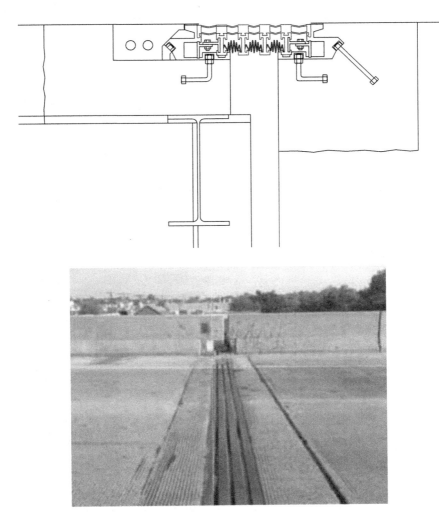

Figure 5.38 *Modular Joint Expansion Joint.*

deterioration of the device and the surrounding concrete, or the previous joint may have been undersized, or the previous joint design may have proved lacking. Selection of the new joints is dependent on three points:

1. The amount of thermal movement to be accommodated;
2. If the device is placed along a skew relative to the direction of thermal movement, or if the superstructure is curved; and
3. The depth of space available to construct and attach the device.

Different joints types are suitable for different ranges of movement. This is in part because of limitation of movement in glands and, in part, because

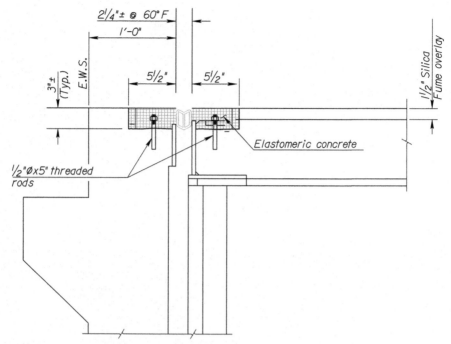

Figure 5.39 *Preformed Neoprene Gland Expansion Joint.*

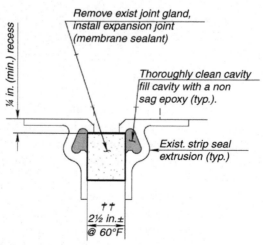

Figure 5.40 *Replace Strip Seal Gland With Membrane Sealant.*

of the need to limit the maximum gap at the joint to 4 in. The following joint types are suitable up to the maximum movement as shown here:

- Compression seals   up to 2 in.
- Neoprene seals   up to 3 in.
- Strip seals   up to 4 in.
- Finger plates, sliding plates, and modular devices   over 4 in.

Although finger plates have been successfully used on skewed and curved bridges, there is potential for misaligning the fingers during installation. Also, fingers are not tolerant of movement that is not parallel to the orientation of the fingers. Such movement may be encountered on curved bridges due to the differing lengths of the girders.

Although modular devices can accommodate large movements and complex geometry, they require more depth than the alternative joints to house supporting structure under the extrusions and glands. Their complexity also makes them more expensive in both initial construction and maintenance costs (Table 5.1).

If the thermal movement at the abutment is 2.5 in. or less, elimination of the expansion joints in the deck could be considered. For a symmetrical bridge fixed against longitudinal movement by bolsters at a center pier this is the movement expected from a 400-ft.-long steel superstructure. This is done by forming the abutment backwall continuously with the deck. If the abutment backwall and deck are monolithic with the abutment beam, the abutment is referred to as an integral abutment. If not, it is a semi–integral abutment. The foundation elements of an integral abutment may be able to accommodate the lateral movement imposed by thermal expansion and

Table 5.1 Expansion joint characteristics for selection

Joint type	Maximum movement (in.)	Expected life (years)	Other notes
Compression seal	2	5	
Neoprene seal	3	10	
Strip seal	4	15	
Finger plate	>4	50	May have trouble with skewed or curved superstructures
Sliding plate	>4	50	
Modular joint	>4	20	Requires sufficient depth for supporting mechanisms

contraction of the superstructure. For this reason, integral abutments are not suited to abutments on timber piling or those founded on footings in rock. Typically, only bridges with stub abutments on steel piling are suitable candidates for reconstruction of the abutments as integral.

When analyzing existing pile groups for the loading imposed by thermal movement of the superstructure, it must be remembered to take into account the effects of skew. Skew angles increase the effective stiffness of pile groups considerably, which will result in higher moments being imposed on the pile by lateral movement. Many bridge owners limit the use of integral abutments to bridges with skew angles of 30° or less.

Lateral movement of the piling may also effect retaining walls adjacent to the abutment, particularly Mechanically Stabilized Earth (MSE) walls which depend on the reinforced fill behind the face of wall for stability. For this reason, a retrofit with an integral abutment may not be practical in locations with MSE walls in the bridge berm, as found at many grade separations.

A semi-integral abutment layout isolates the foundation elements of the abutment from the effects of thermal movement of the superstructure. To ensure this, detailing of the abutment beam reconstruction should include a gap between the backwall and the beam (Figure 5.41).

Neither an integral nor a semi-integral abutment eliminates the thermal movement of the superstructure. The movement is simply carried through to the approach. Either layout should include a concrete approach slab with a joint capable of dealing with the required thermal movement.

5.4.3 Bearing Devices

Bearings serve to isolate the substructure from the rotation of the superstructure elements in flexure. They are typically found on steel bridges and at unit breaks on prestressed and posttensioned concrete bridges. As discussed in Section 2.3.2.4, the bearing types are fixed (accommodating rotation only) and expansion (accommodating rotation and translation). The materials most commonly used are steel and elastomer (Figures 5.42 and 5.43).

Steel bearings are subject to corrosion, particularly when expansion joints fail and allow water to drain onto the bearing seat. Corrosion product can lock the bearing into place, resulting in unintended loading being introduced into the substructure and unintended restraint exerted on the superstructure. Because the corrosion product of steel has a greater volume than the uncorroded steel, corrosion at the bearing can force the superstructure upward, resulting in a grade differential across expansion joint in the deck above.

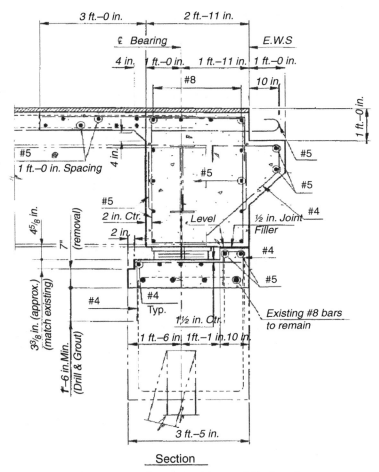

Figure 5.41 *Backwall Reconstruction as Semi-Integral Abutment.*

The minimal substantial maintenance for corroded steel bearings is to clean and paint them. If there has been loss of section, or if mating surfaces are simply too heavily pitted, components of the bearing device replaced. Severely corroded bearings may be replaced, entirely (Figure 5.44).

Elastomeric bearings do not rust, but the material can deteriorate, particularly if loads encountered in service are greater than those assumed in design or if the fabrication of the bearing did not meet specifications. Bearings with reinforced elastomeric pads may be replaced if the material is split or severely cracked (Figure 5.45).

Elastomeric bearings which must accommodate a significant amount of translation use a sheet of polytetrafluoroethylene (PTFE, best known

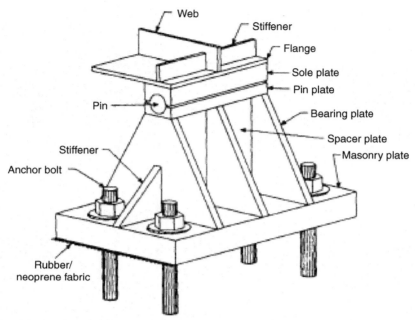

Figure 5.42 *Parts of a Steel Bolster. (Adapted from Federal Highway Administration. Bridge maintenance training. Washington, DC; US Department of Transportation; 2003).*

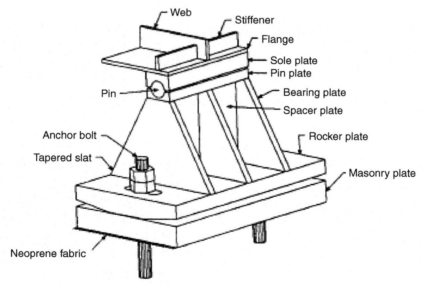

Figure 5.43 *Parts of a Steel Rocker. (Adapted from Federal Highway Administration. Bridge maintenance training. Washington, DC; US Department of Transportation; 2003).*

Figure 5.44 *Deteriorated Steel Rocker Bearing.*

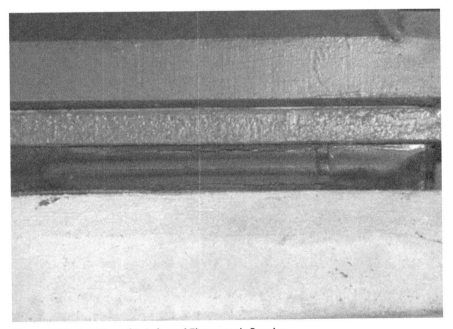

Figure 5.45 *Splitting of Reinforced Elastomeric Bearing.*

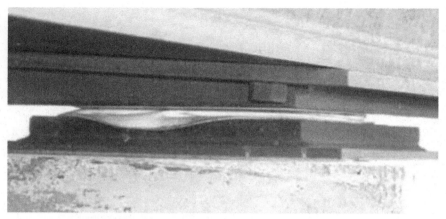

Figure 5.46 *Teflon Sliding From PTFE Bearing.*

by the brand name Teflon™). This sheet may "walk out" if not properly set. Maintenance actions may be needed to reset the PTFE sheet (Figure 5.46).

More complex bearings used for high-bearing loads and for complex movements resulting from curvature of the superstructure, such as pot and disc bearings, may have component failure. However, it is typically less expensive to replace such bearing rather than repair the device in place.

Movement of substructure elements resulting from soil pressure, foundation settlement, or thermal movement transmitted through corroded bearing, may result in bearing falling out of their initial alignment. Bearings may reset as long as there is sufficient room on the bearing seat to facilitate the adjusted bearing location.

All of the earlier-mentioned scopes of work, except for simple cleaning and painting, will require the jacking of the superstructure to remove the load from the existing bearing to allow it to be moved. A jacking plan for a highway bridge in service should be developed by a competent, licensed engineer. In lieu of a more precise structural analysis of an existing superstructure, for a steel bridge the jacking loads may be estimated from the bearing capacity of the existing girder web and bearing stiffener at the location of interest using the formula:

$$R = \min(19A_G, 25A_C) \quad \text{For 36 ksi steel}$$

Where A_G is the gross cross-sectional area of the existing bearing stiffeners and A_C is the cross-sectional area of the bearing stiffeners in contact with the bottom flange of the girder or beam (Figure 5.47).

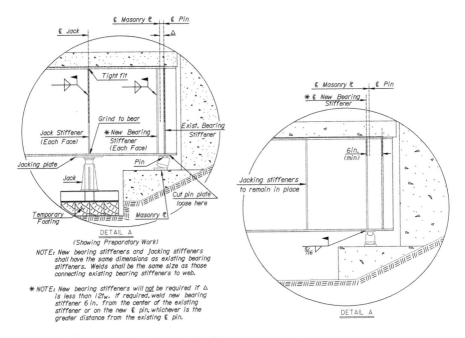

Figure 5.47 *Steel Bearing Removal and Reset.*

If the existing bearings are corroded, the bearing seat area typically suffers from deterioration. Damaged areas of concrete should be removed and replaced during the substantial maintenance action. Consideration should be made of the location of damaged areas when sequencing any jacking, if the bearing seat is intended to be used as support for the jacks.

5.4.4 Steel Superstructure

Most of the typical repairs to steel superstructures were discussed in Section 5.3.2 in the review of repairs for damage induced by fatigue, corrosion, and impact were discussed. These repairs constitute the bulk of substantial maintenance actions required for steel superstructures. Four other actions discussed include: gusset plate repair, the repair or replacement of truss members, strengthening beams and girders, and repair or replacement of pin and hanger devices.

Trusses constitute an ever-shrinking part of the bridge inventory in the United States. In the early part of the twentieth century labor was cheap, material costs were relatively high, and the transport of large pieces of steel from the fabrication shop to the job site was difficult. This made truss superstructures an attractive option for short and moderate span length bridges.

The inversion of the ratio of material to labor cost and the improvement in the nation's network of highways in the latter half of the twentieth century changed the economics of steel bridges to favoring beam and girder superstructures. Also contributing to the decline in the use of trusses is the vulnerability of the structure type to corrosion.

One aspect of substantial maintenance on trusses that is given far more attention now than it was in previous year is the evaluation and repair of gusset plates. It was the failure of an undersized gusset plate connection that caused the collapse of the I-35W Bridge in Minneapolis, Minnesota in 2007. Prior to that, the bridge engineering community in the United States had assumed that the design assumptions used for sizing gusset plates was conservative enough to assure that other truss members would fail before the gussets would.

Gusset plates, particularly on the lower chords, receive a great deal of exposure to moisture and road salts, which may sit in the corners and crevices of the connection. The moisture and salt will continue to drive corrosion after the rest the superstructure has dried. Any loss of section identified in bridge inspections should be evaluated by the methods discussed in the FHWA publication, *Load Rating Guidance and Examples for Bolted and Riveted Gusset Plates in Truss Bridges* [11]; and the AASHTO *Manual for Bridge Evaluation* [12]. If repair is required, it will typically consist of adding steel section across the predicted failure plane and replacing rivet with high-strength bolts (Figures 5.48 and 5.49).

Trusses may also require the addition of section to existing built-up members due to either losses from corrosion or lack of capacity in the original design. Because of the age of most trusses, if welding is used when adding section to a member, great care should be taken to ascertain that the existing material is weldable. As discussed in Section 2.5.3.2, steels with high-carbon equivalent should not be welded. Subsequent fracture of the weld and the existing steel may result (Figure 5.50).

In a similar manner, deficient steel beams and girders may be strengthened with the addition of steel section. Cover plates may be attached to the bottom flange to increase the positive moment capacity of a steel beam or girder. Note, that as with any addition of section to gain capacity after a member is in service; any dead load stresses from originally being carried by a member will remain in the existing member section unless those stresses are relieved (possibly by jacking the member) and then reintroduced to the new augmented section.

One strengthening method that can be used to relieve existing stresses on a member is post-tensioning (Figure 5.51). For a beam or girder, the most common method is to attach post-tensioning rods to the bottom

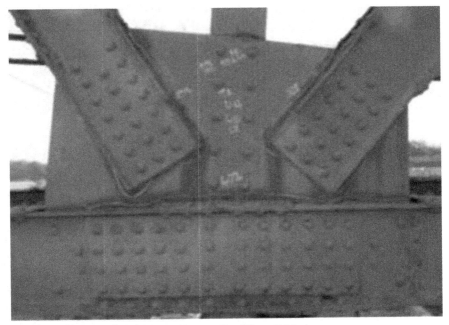

Figure 5.48 *Deteriorated Gusset Plate on Steel Truss Bridge.*

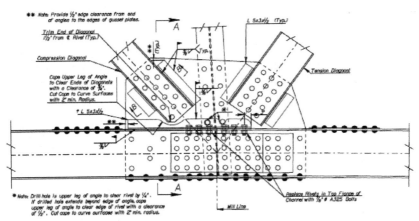

Figure 5.49 *Repair for Deteriorated Gusset Plate.*

flange. This method adds strength, reduces deflection, and can reduce tensile stresses contributing to fatigue.

A detail which was previously in general use in the design of long span girders, but which has fallen out of favor is the pin and hanger joint. Pin and hanger joints are used to suspend the interior section of a span from

Added metal	Description	Figure
Cover plate/ plates	Cover plate is welded to the compression member to increase the steel area and reduce the slenderness ratio End post and bottom, chords of a through truss are boxed in with steel plates.	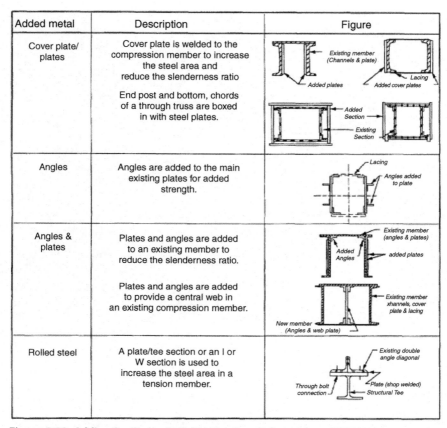
Angles	Angles are added to the main existing plates for added strength.	
Angles & plates	Plates and angles are added to an existing member to reduce the slenderness ratio. Plates and angles are added to provide a central web in an existing compression member.	
Rolled steel	A plate/tee section or an I or W section is used to increase the steel area in a tension member.	

Figure 5.50 *Adding Section to Built-Up Member.* (Adapted from AFJAPM 32-1088 Bridge Inspection, Maintenance and Repair [12]).

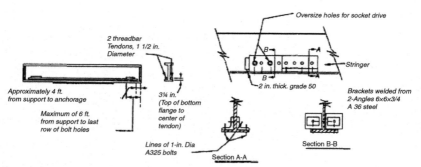

Figure 5.51 *Posttensioning of a Steel Beam.* (Adapted from AFJAPM 32-1088 Bridge Inspection, Maintenance and Repair [13]).

cantilever spans over piers at either end. These designs were used in the past to simply analyze the girder, to facilitate steel erection before bolted splices with high-strength bolts were widely used, and to move the deck expansion joints away from bearing locations at piers. However, the failure of either the pin or hanger, can lead to a catastrophic collapse, as seen with the collapse of the I-95 Bridge over the Mianus River in Greenwich, Connecticut in 1983.

That collapse led to the implementation of fracture critical bridge inspections in the United States. When a defect is found in a pin and hanger, there are three options for repair. First, the joint may be kept but the pins and/or hangers replaced. Rather than replacing strictly in kind, the material used is upgraded. Pin and hanger devices in the 1950s were often fabricated with mild steel. New pins are typically stainless steel and new hangers are fabricated from modern, fracture resistant steels. The fit of the pin is also much tighter, which reduces friction and the infiltration of moisture.

A second option is to remove the joint, make the girder continuous at that location, and relocate the point of expansion and contraction to the ends of the bridge. Note, typically there would be no existing pier wide enough to facilitate the two sets of bearing required for a unit break. This is contingent on the girder having sufficient capacity in its new configuration.

A third option is to replace the pin and hanger with a more robust joint configuration, such as a shelf (Figure 5.52). In this configuration, the suspended span rests on an elastomeric bearing device on a shelf protruding from the cantilever span. This does not change the configuration of the girder for moment distribution, but provides a joint that has better access for inspection and is less vulnerable to corrosion.

Figure 5.52 *Replacement of Pin and Hanger Joint With Shelf.*

5.4.5 Painting Steel Structures

Protective coatings act to prevent corrosion of steel by providing a barrier against moisture, oxygen, and other contaminants; and/or by providing galvanic action. Paint is the most common protective coating applied. Galvanizing is more limited in use on bridge structures and is typically limited to particular elements, such as cables or components of expansion joints that require a high degree of protection and are small enough to be hot dipped. It is not applied to existing bridges as a maintenance action.

Although painting would appear to belong to the realm of preventative maintenance, the high cost and the amount of work required to prepare and repaint an existing bridge relegates it to substantial maintenance. Because of the cost, paint could be considered as a separate element of the bridge for tracking in bridge management systems.

Much of the cost of repainting is attributable to the cost of removing existing paint systems and properly preparing the surface. Most paint systems require the preparation of the surface to be completely free of any oil, mill scale, corrosion products of other contaminates in any 3×3 in. area over 95% of the surface. This is referred to as near-white, as that the prepared surface should shine, and must remain free of contaminates until the new paint is applied. Some more tolerant paint systems utilizing organic primers may only require a commercial-blast quality of preparation, which requires 66.7% of surface in any 3×3 in. area to be completely free from contamination (Figure 5.53).

The cost and effort of removal is considerably more if the existing paint system contains lead. As discussed in Section 4.5, a lead content higher than 5 mg/L requires that the wasted paint material be dealt with as hazardous waste. During the removal process all wasted paint, including the airborne particulates must be contained and kept out of the surrounding environment.

A number of paint systems for bridges exist with differing systems favored by different bridge owners. For repainting, it is recommended that the primer used be a zinc-rich system. This primer not only provides a barrier, but acts as a sacrificial anode to protect the steel. Overcoating systems, as discussed for spot painting, are not recommended for complete bridge repainting unless required by extenuating circumstances that would preclude complete removal of the existing paint. The paint applied during overcoating depends on the adhesion of the existing system to the steel. Additionally, not removing the existing paint system precludes the complete removal of existing contaminants from the steel surface.

If repainting of the entire bridges is too expensive, consideration can be given to zone painting. The areas near expansion joints and the exterior

Figure 5.53 *Containment for Paint Removal on Steel Girder Bridge.*

fascia of the exterior girders are the areas where a protective system is most challenged. It is also not uncommon for the existing paint system outside of these zones to be in good condition while failing elsewhere on the bridge. Particularly for substantial maintenance work involving replacement of expansion joints, if the existing system does not contain lead, repainting of the first few feet of girder adjacent to the joint should be considered.

Zone painting at expansion joints should also be considered even if the existing steel is weathering steel. Weathering steel forms a thick protective patina that protects the interior of the cross-section from further corrosion. It is common in newer construction. However, the same propensity of the material to form a patina can continue unabated if the weathering steel is in an environment with continually high moisture, as might be encountered under a failed expansion joint. For this reason, weathering steel should never be used in the fabrication of bearing devices, expansion devices or steel grid decking. Many devices so fabricated and installed in the 1980s failed in the subsequent decade.

5.4.6 Reinforced Concrete Superstructure

The majority of repairs to reinforced concrete superstructure elements involve sealing and injecting cracks and concrete surface repair, as discussed in Section 5.3.1. However, substantial maintenance actions may be required to address deficiencies in capacity that exist from the original construction of a bridge member. A particular problem exists with reinforced concrete deck girders designed in the late 1940s and in the 1950s. The AASHO design specification at the time increased the allowable shear stress of concrete and reduced the required development length for reinforcing steel (by allowing a higher bond stress). This resulted in overestimating the shear capacity of the reinforced concrete deck girders. In the decades since, multiple states have had to add shear capacity to these bridges in service.

For beams and girders, members dominated by flexure, additional shear capacity may be provided with additional reinforcing steel (Figure 5.54). A typical repair for reinforced concrete deck girder is to drill and grout reinforcing steel into the member. This is most effective if the reinforcing was placed normal to the direction of cracking.

Similar to its use as a retrofit for steel superstructures, post-tensioning can be used to introduce compressive stresses, counter deflections, and increase flexural capacity of reinforced concrete members (Figure 5.55).

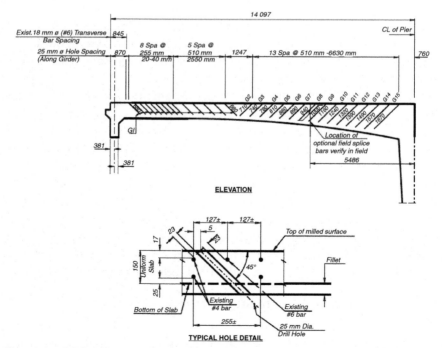

Figure 5.54 *Rebar Insertion for Shear Strengthening of Reinforced Concrete Girder.*

Figure 5.55 *Posttension Retrofit of Pier Cap.*

The method that has come to be preferred for adding shear and/or flex-ural capacity to reinforced concrete members is the use of FRP reinforcing (Figure 5.56). Note, that most bridge owners prefer carbon fiber material,

Figure 5.56 *Installation of FRP Repair on Prestress Concrete Beam.*

rather than glass fibers, for structural uses due to its superior strength. Using externally bonded FRP reinforcing strips provides for a significant increase in strength without the need for heavy construction equipment or prolonged construction times. Work in the field to apply FRP consists of concrete surface repair of any damaged concrete at the location of attachment for the FRP to provide a sound substrate, followed by a final cleaning, applying the reinforcing strips, and then the application of any protective coat to seal out ultraviolet light from FRP material.

FRP systems are proprietary in nature, however, there is a guide specification for the design of these repairs in NCHRP Report 655 *Recommended Guide Specification for the Design of Externally Bonded FRP Systems for the Repair and Strengthening of Concrete Bridge Elements* [14]. The reader is referred to the report for the specifics of designing such repairs; however, the report authors note that designs should have the following attributes:

1. Flexural reinforcing schemes should be checked to ensure that the existing steel tension reinforcement in the member will yield before the FRP debonds from the surface. This is to ensure a ductile failure in flexure in the member.
2. Shear reinforcing should wrap around the member as much as practicable to fully develop the FRP. Anchors may be used where this is not possible.
3. If practical, 1 in. gaps should be left between adjacent strips to allow for future inspection of the adhesion of the FRP to the substrate.

5.4.7 Prestressed Concrete Superstructure

The average age of the nation's stock of 144,000 prestressed concrete bridges is 30 years old. This is significantly younger than the average age of 44 years for all bridges in the National Bridge Inventory. Given this, and the advantage of having members precompressed to keep cracks closed, it would be expected that there are fewer substantial maintenance issues for these types of bridges than for others in the inventory. However, these bridges have not proved impervious to deterioration or the effects of corrosion.

On December 28, 2005, an exterior prestressed concrete box beam of the Lake View Drive Bridge in South Strabane, Pennsylvania collapsed onto Interstate 70 below. The beam failed under dead weight. There was no traffic on the bridge at the time. The cause was attributed to corrosion of the strands [15].

Prestressed concrete bridges may suffer from the effects of corrosion; however, these bridges may show fewer indications of deterioration and loss of capacity prior to failure than steel or reinforced concrete superstructure bridges. A sudden loss of prestressing does not provide for a ductile failure.

The repair for damaged concrete in a prestressed beam is the same as it is for conventionally reinforced concrete, crack sealing, and concrete surface repair. The same care that should be taken with these activities for reinforced concrete is even more important with prestressed concrete. When removing damaged concrete, care must to taken to not damage or nick the prestressing strands. Also, when patching material is selected, consideration must be made to avoid the halo effect of accelerating corrosion in the surrounding existing concrete.

The repair impact damage to prestressed concrete beams add another factor to deal with, the repair of severed strands. When the bottom strands are severed in a prestress beam, there is not only a loss of flexural capacity, but also of camber. Losing over 5% of the strands can be expected to result in a complete loss of camber. More than that and the beam can be expected to deflect downward under dead load. When repairing the beam, the prestressing lost from the damaged strands needs to be restored. Three viable options for restoring the tensioning are:

1. Splicing the severed ends of the strands with mechanical couplers that act as turnbuckles;
2. Attaching external posttensioning rods; and
3. Externally bonding tensioned carbon-FRP strips to the beam.

Mechanically splicing the severed strands is a common repair suitable for most repairs (Figure 5.57). These couplers are significantly larger in diameter than the prestressing strands; however, and the repair of multiple adjacent strands requires staggering the splices. Although this is typically not a problem for I sections, it can be for box sections. Another limitation is the availability of couplers for only up to 0.5 in. diameter strands (larger prestressed concrete members are beginning to use 0.7 in. diameter strands).

To help control cracking and to provide additional shear capacity, vertical wraps of FRP may be applied to the repaired area. As with the repairs to reinforced concrete, the wrap should go around the perimeter of the beam.

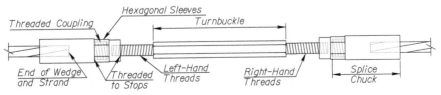

Figure 5.57 *Mechanical Splice for Prestressing Strand.*

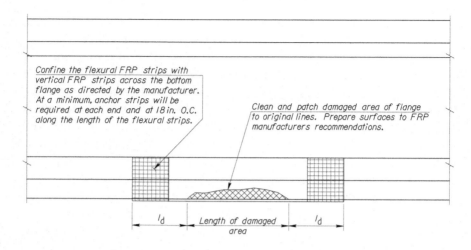

Confine the flexural FRP strips with vertical FRP strips across the bottom flange as directed by the manufacturer. At a minimum, anchor strips will be required at each end and at 18 in. O.C. along the length of the flexural strips.

Clean and patch damaged area of flange to original lines. Prepare surfaces to FRP manufacturers recommendations.

l_d Length of damaged area l_d

TYPICAL FRP FLEXURAL STRENGTHENING OF PRESTRESSED CONCRETE GIRDERS

Figure 5.58 *Prestressed Beam Repair With FRP.*

To control cracking by inducing compression in the patched area for the dead load only condition, many states require preloading the beam during placement of the new concrete. Preloading is applying a vertical force to the beam (Figure 5.58). Commonly this is provided by a truck parked over the beam, inducing a download deflection in the beam during patching. When the truck is removed, the patched area is compressed with the upward rebound of the beam.

The final report for NCHRP Project 20-07 Task 307 includes the document, *Guide to Recommended Practice for the Repair of Impact-Damaged Prestressed Concrete Bridge Girders* [16]. This document contains guidelines for evaluating and repairing damage and shows the required calculations.

5.4.8 Posttensioned Concrete Superstructure

The main concern for the maintenance of post-tension concrete bridges is maintaining the integrity of the post-tension strands. Corrosion of the high-strength steel strands can occur when they are exposed to water and chlorides. Corrosion of post-tension strands, and subsequent bridge failure, has resulted from exposure of the strands due to failure of the protective systems for the strands. Bonded post-tensioning systems, which make up the large majority of post-tensioned concrete superstructures, protect the strands within metal or plastic ducts with a cement grout. Voids in the grout may result from construction errors, compounded by insufficient inspection, during the initial grouting; or from segregation or shrinkage of the grout. Bleed water from the grout, or water traveling through leaks in the ducts, can drive corrosion in these voids. If the source of water is external, chlorides may be introduced from road salts, or from marine environments. More troublesome for the industry, from 2002–2010, a major supplier of grout for post-tensioning had produced grouts with potentially elevated chloride levels [17].

In 2013, the FHWA issued the publication, Guidelines for Sampling, Assessing, and Restoring Defective Grout in Prestressed Concrete Bridge Post-Tensioning Ducts [18]. It provides direction on how to conduct a physical investigation of post-tensioning ducts, and on the sampling of in place grout and subsequent material investigation. Using Non-Destructive Testing (NDT) methods such as ultrasound and ground-penetrating radar, post-tensioning ducts are located. A small area of the concrete cover is removed and the top of the duct is drilled into. The interior is inspected for voids and grout samples are taken. If a void is present, the hole may be capped to retain access for subsequent regrouting (Figure 5.59).

5.4.9 Piers and Pier Bents

The vast majority of highway bridge piers are constructed from reinforced concrete. The vast majority of structural maintenance work on these piers consist of concrete surface repair. Concrete in piers is subject to damage from corrosion, from vehicular impact and from fire. The methods of concrete surface repair are discussed in Section 5.3.1.

As it is with reinforced concrete superstructures, the preferred method for strengthening members of reinforced concrete pier is externally bonded FRP. FRP may be applied to pier beams in the same manner as to superstructure

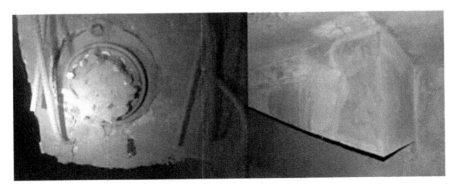

Figure 5.59 *Repair of Posttensioning Anchorage.*

beams. FRP may be applied to columns to add strength by increasing confinement. An increase in confinement, for columns with an effective length to diameter ratio of eight or less, allows use of a higher value of compressive strength in the analysis of the member [14] (Figure 5.60).

The leading cause of bridge collapse in the United States is not deterioration of reinforced concrete, but rather it is scour. As the substructure element in the channel, a major focus of attention for substantial maintenance should be the protection of piers from the effects of scour. The primary method to protect the foundations of bridges in-service from scour is armoring, that is, typically the placement of heavy rock around the foundation in the channel.

Although it may seem a simple matter to dump rock adjacent to a pier, the proper installation of armoring does require engineering and monitoring on subsequent inspections. The pier of the I-90 Bridge over Schoharie Creek near Albany, New York, whose collapse led to the implementation of underwater inspection requirement in the United States, were armored with rock riprap during their original construction. However, the rock riprap had washed downstream over the 34 years that the bridge was in service.

Hydraulic Engineering Circular #23 (HEC-23), *Bridge Scour and Stream Instability Countermeasures Experience, Selection and Design Guidance* by the FHWA discusses the selection and design of various hydraulic countermeasures for river training and for armoring to protect bridge piers and abutments [19]. River training countermeasures are intended to alter the flow and velocity of water to mitigate adverse erosion or deposition of streambed material along a reach of the channel. Armoring protects the streambed from hydraulic stresses, and therefore, erosion. Other than rock riprap, there are other options for armoring at piers. They are typically shaped concrete products (like A-Jacks™) with greater capacity for interlocking than does rock, or bags of cement or grout which may be more convenient to locate and handle than large diameter rock. The majority of armoring used, though, is rock riprap (Figure 5.61).

Key to designing the rock armoring is selection of the proper size rock necessary to resist hydraulic shear force for a given flow. HEC-23 provides the following equation for selection of the necessary median size of rock.

$$D_{50} = \frac{0.692\ (KV)^2}{(S_s - 1)2g}$$

Where, D_{50}, median stone diameter in ft.; K, coefficient for pier shape (1.5 for round nosed piers, 1.7 for rectangular piers); V, channel velocity at the pier in ft./s; S_s, specific gravity of riprap (typically 2.65); and g, gravitational acceleration constant (32.2 ft.²/s).

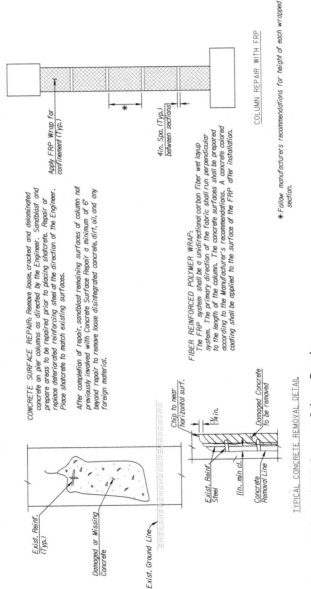

CONCRETE SURFACE REPAIR: Remove loose, cracked and delaminated concrete on pier columns as directed by the Engineer. Sandblast and prepare areas to be repaired prior to placing shotcrete. Repair or replace deteriorated reinforcing steel at the direction of the Engineer. Place shotcrete to match existing surfaces.

After completion of repair, sandblast remaining surfaces of column not previously involved with Concrete Surface Repair a minimum of 6" beyond repair to remove loose disintegrated concrete, dirt, oil, and any foreign material.

FIBER REINFORCED POLYMER WRAP:
The FRP system shall be a unidirectional carbon fiber wet layup system. The primary direction of the fabric shall run perpendicular to the length of the column. The concrete surfaces shall be prepared according to the Manufacturer's recommendations. A concrete colored coating shall be applied to the surface of the FRP after installation.

Apply FRP Wrap for confinement (Typ.)

4 in. Spa. (Typ.)
between sections

COLUMN REPAIR WITH FRP

* Follow manufacturer's recommendations for height of each wrapped section.

Exist. Reinf.
(Typ.)

Damaged or Missing
Concrete

Exist. Ground Line

Chip to near
horizontal surf.

¾ in.

Exist. Reinf.
Steel

1in. min cl.

Concrete
Removal Line

Damaged Concrete
to be removed

TYPICAL CONCRETE REMOVAL DETAIL

Figure 5.60 Typical FRP-Concrete Column Repair.

Figure 5.61 *Placing Rock Around Pier in Missouri River.*

The width of the mat of riprap should extend upstream and down-stream a minimum of twice the pier width parallel to the direction of flow. The top of the mat should be the same elevation as the streambed in normal flow. Burying the mat may hide it from bridge inspectors, while protruding

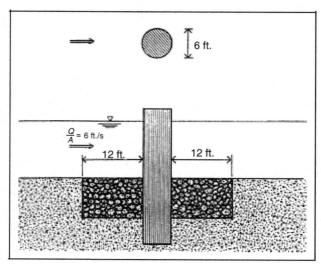

Figure 5.62 *Placement of Rock Riprap About Pier Column. (Adapted from HEC-23 [19]).*

above the streambed encourages turbulence, which may lift the riprap and carry it downstream during high flow events. The thickness of the riprap mat should be three times D_{50} or more. The bottom of the mat should be at or below the estimated depth of contraction scour at the pier. The rock used should be well graded with no rock larger than twice D_{50} (Figure 5.62).

A layer of geotextile or gravel filter material place below the riprap mat may improve performance by relieving hydrostatic pressure under the mat. The same effect is achieved to some degree by using well-graded rock for the riprap. Partially grouting the riprap will reduce the size of the rock needed, and is an option where large rock is not available. Fully grouting the riprap is not recommended as it provides no relief for hydrostatic pressure under the continuous mass.

5.4.10 Abutments

Most highway bridge abutments are constructed from reinforced concrete. As with piers, the most common substantial maintenance action addressing structural concerns at abutments is concrete surface repair. There are two issues that affect abutments and not piers, drainage from the approach roadway and earth pressure.

Uncontrolled drainage from the approach may wash around abutment wings and underneath stub abutments, leaving the foundations exposed. Improving drainage at approach will help mitigate this (Figure 5.63). Another option is to excavate the fill and install a drainage system adjacent to

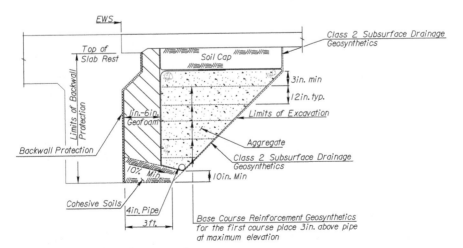

Figure 5.63 *Abutment Drainage System.*

the abutment. This may by a proprietary backwall drain system or a section of drained aggregate from which water is piped away. Excavation may also allow for the installation of granular backfill material to relieve excess pressure on the abutment.

Scour is an issue that must be addressed at abutments, particularly for shorter span structures where the setback of the abutment from the channel may be small. Scouring action is more complex at abutment than at piers due to vertical flows and vortices that develop at the approach embankment.

To size the median rock required for riprap, HEC-23 requires the calculation of the Froude number.

$$Fr = \frac{V}{\sqrt{gy}}$$

Where, Fr, Froude number; y, depth of flow in ft. at the floodplain adjacent the approach embankment; and g, gravitational acceleration constant.

For Fr equal to or less than 0.80, the median diameter of rock is given by:

$$D_{50} = \left(\frac{K}{(S_s - 1)}\right)\left(\frac{V^2}{g}\right)$$

For Fr greater than 0.80, the median diameter of rock is given by:

$$D_{50} = y\left(\frac{K}{(S_s - 1)}\right)\left(\frac{V^2}{gy}\right)^{0.14}$$

Where, D_{50}, median stone diameter in ft.; V, average velocity in the contracted section (flow above the floodplain) in ft./s; and S_s, specific gravity of riprap (typically 2.65).

The coefficient, K, varies for type of abutment and Froude number (Table 5.2).

Table 5.2 K value for determining rock size at abutment

	Spill through abutments	Vertical wall abutments
Fr ≤ 0.80	0.89	1.02
Fr > 0.80	0.61	0.69

The rock protection should extend across the entire length of the toe of the abutment and should extend out from the toe into the waterway a distance of twice the flow depth, y (Figures 5.64 and 5.65).

5.4.11 Culverts

Substantial maintenance work for culverts mostly consist of types already covered, concrete surface repair and (in Chapter 4) mudjacking subsurface voids. Just as with abutments, culvert wings face challenges from earth pressure and poor drainage from the roadway above. However, it is typically

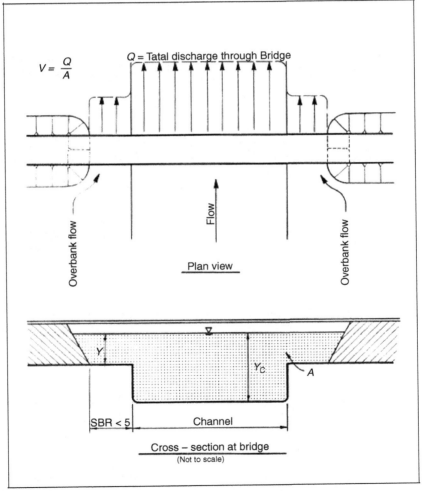

Figure 5.64 *Cross-Section at Bridge Opening. (Adapted from HEC-23 [19]).*

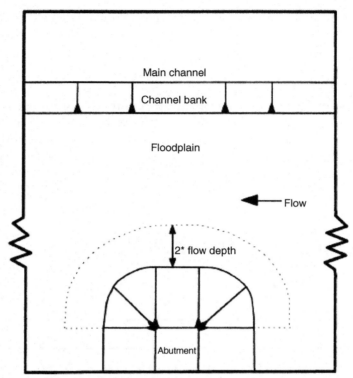

Figure 5.65 *Plan View Showing Extend of Riprap at Abutment. (Adapted from HEC-23 [19]).*

cheaper to replace damaged wings rather mitigate with option such as shoring.

A substantial maintenance action particularly applicable to culvert is lining. For corrugated metal pipe culverts, acidity in the soil and failure of galvanizing system can lead to section loss. While excavation and replacement is an option, this necessitates traffic disruption. If the hydraulic opening is sufficient that either a concrete or a plastic pipe with conveyance equal to the original opening (due to the lower hydraulic roughness of the liner material) will fit inside, then placing the new pipe and filling the void with concrete is an attractive option.

For reinforced concrete culverts, repair of the deteriorated concrete by patching is problematic if the deterioration extends completely through the section (Figure 5.66). In this case, lining is done by construction of a corrugated metal arch inside the opening and the annular void is filled with grout.

Figure 5.66 *Reinforced Concrete Arch Culvert Lined With Corrugated Steel.*

5.5 REHABILITATION ACTIONS

Substantial maintenance actions are intended to improve or maintain the condition of a bridge sufficiently to allow it to remain fully functional. Sometimes the functionality provided by the existing bridge's roadway width, or vertical alignment; or its load-carrying capacity is insufficient for current needs. Construction intended to improve the functional characteristics of a bridge to meet current requirements is rehabilitation. The most common rehabilitation action on highway bridge is deck replacement, often referred to as redecking.

5.5.1 Deck Replacement

The most common rehabilitation action on highway bridges is deck replacement, often referred to as redecking. Redecking a bridge may be the least expensive option for beam or girder bridges with a high percentage of deterioration in the existing deck. Redeck allows the following improvements to be made:

• Upgrading bridge railing to current standards;
• Upgrading bridge deck expansion joints and better protecting other elements from drainage;
• Small adjustments in cross-slope and in vertical alignment to improve ride;

- Slight increase in roadway width (contingent on the existing girder spacing and capacity); and
- An increase in load–carrying capacity if the new deck is made composite with the girders.

Often the condition or characteristics of the existing bridge are such that it is not possible to bring it up to full current design standards. However, in the United States, most states implemented 3R Standards in response to the Surface Transportation Assistance Act of 1982. The 3Rs are resurface, rehabilitate, and reconstruct. The intent behind the implementation of these standards was to improve the safety of nonfreeway bridges by substantially improving features on a large portion of the nation's road and bridge inventories economically. Each state adopted its own 3R policies and cooperation with its state FHWA office. In most states 3R standards are sufficient for redecking and other rehabilitation projects on bridges not on freeways or the Interstate system.

It is common to make replacement deck acts compositely with existing steel girders with the addition of headed steel studs to the top flange. This also reduces live load deflections in the rehabilitated structure (Figure 5.67).

Figure 5.67 *Deck Forming for Redeck Project.*

5.5.2 Superstructure Replacement

If the existing superstructure is in poor shape, or if a large change in vertical alignment is required, then the entire superstructure may be replaced. Typically, a steel girder superstructure would be replaced with a new steel girder structure; and, it is similar for prestressed concrete beams. The span arrangements are made in the original design to optimize the use of one material or the other. In case where it is being considered to replace an existing steel superstructure with prestressed concrete, consideration must be made at the earliest stages about the impact of the significantly greater weight of prestressed concrete structures on the existing foundations.

5.5.3 Bridge Widening

If the current roadway of a bridge is insufficient, the bridge deck may be replaced at the structure widened with the addition of substructure and superstructure. This is a more common scenario when a highway facility such as an interchange is being upgraded rather than a site-specific bridge project.

REFERENCES

[1] American Concrete Institute Committee 224. Causes, evaluation, and repair of cracks in concrete structures. Farmington Hills, MI: American Concrete Institute; 2007. ACI 224.1R-07.
[2] Ball JC, Whitmore DW. Corrosion Mitigation Systems for Concrete Structures. Concrete Repair Bull 2003;6–11.
[3] Dexter RJ, Ocel JM. Manual for repair and retrofit of fatigue cracks in steel bridges. Washington, DC: Federal Highway Adminstration; 2013. FHWA-IF-13-020.
[4] Avent RR, Mukai D. Heat-straightening repairs of damaged steel bridges. A manual of practice and technical guide. Washington, DC: Federal Highway Administration; 1998. FHWA-SA-99-034.
[5] Anderson B, et al. Post-retrofit analysis of the Tuttle Creek Bridge. Topeka, KS: Kansas Department of Transportation; 2008. K-TRAN: KU-06-2.
[6] Bennett C, et al. Enhancement of welded steel bridge girders susceptible to distortion-induced fatigue. Lawrence, KS: University of Kansas Center for Research, Inc; 2012. TPF-5(189).
[7] Mishler HW, Leis BN. Evaluation of repair techniques for damaged steel bridge members phase I. Washington, DC: Transportation Research Board; 1981. NCHRP Report 12–17.
[8] Shanafelt GO, Horn WB. Guidelines for evaluation and repair of damaged steel bridge members. Washington, DC: Transportation Research Board; 1984. NCHRP Report 271.
[9] Connor RJ, Kaufmann EJ. Urban MJ. Heat-straightening repair of damaged steel bridge girders: fatigue and fracture performance. Washington, DC: Transportation Research Board, National Research Council; 2008. NCHRP Report No. 604.
[10] Wright W, et al. Highway bridge fire hazard assessment draft final report. Blacksburg, VA: Virginia Polytechnic Institute and State University; 2013. NCHRP Project No. 12-85.

[11] Load rating guidance and examples for bolted and riveted gusset plates in truss bridges. Washington, DC : Federal Highway Administration, 2009. FHWA-IF-09-014.

[12] American Association of State Highway and Transportation Officials. The manual for bridge evaluation. 2nd. ed. Washington, DC : s.n.; 2011. MBE-2.

[13] Bridge Inspection, Maintenance and Repair. Washington, DC : Departments of the Army and the Air Force; 1994. TM 5-600/AFJPAM 32-1088.

[14] Zureick A-H, et al. Recommended guide specification for the design of externally bonded FRP systems for the repair and strengthening of concrete bridge elements. Washington, DC: Transportation Research Board; 2010. NCHRP Report 655.

[15] Naito C, et al. Forensic examination of a noncomposite adjacent precast prestressed concrete box beam bridge. Washington, DC : American Society of Civil Engineers; July/August 2010, J. Bridge Eng., pp. 408–418.

[16] Harries KA, et al. Guide to recommended practice for the repair of impact-damaged prestressed concrete bridge girders. Washington, DC: Transportation Research Board; 2012.

[17] Baxter J. Memorandum on elevated chloride levels. Washington, DC: Federal Highway Administration; 2012.

[18] Theyro TS, Hartt WH, Paczkowski P. Guidelines for sampling, assessing, and restoring defective grout in prestressed concrete bridge post-tensioning ducts. Washington, DC: Federal Highway Administration; 2013. FHWA-HRT-13-028.

[19] Lagasse PF, et al. Bridge scour and stream instability countermeasures experience, selection and design guidance. Washington, DC: Federal Highway Administration; 2001. FHWA-NHI 01-003 HEC-23.

CHAPTER 6

Bridge Life Cycle Costing

Overview

The use of bridge life cycle cost analysis (BLCCA) to determine the most cost-effective maintenance strategies is discussed and criteria for selecting and scoping projects are examined. Concepts concerning cash flows and the time value of money are introduced. Costs for inclusion in BLCCA are discussed, as are deterioration rates for bridges and bridge elements. The chapter concludes with an example BLCCA to select a maintenance strategy and discussion on using BLCCA to determine cost-effective maintenance policies.

6.1 PROJECT SCOPING AND SELECTION

Although many tools and techniques have been developed over the past several years to help engineers evaluate bridge maintenance needs on an objective basis, the process of selecting and scoping bridge maintenance projects remains based in engineering judgment. Databases with decades of bridge inspection observations provide the foundations for increasingly sophisticated deterioration models, which allow for more accurate predictions of future bridge conditions. But the selection of maintenance projects is contingent are more than the condition of a particular bridge element. As part of an evolving and ever-changing highway network, which is funded by budgets under increasing demand, the need for a particular project must be vetted by examining a number of criteria.

6.1.1 Initial Selection

The initial selection for bridge maintenance work involves examining four criteria:

1. Is there a deficit that may be economically addressed by maintenance work?
2. Is this bridge necessary?
3. What is the minimum condition necessary for the operation of this bridge?
4. What is the likely remaining service life of this bridge?

These criteria determine if a maintenance action is required, and if so, would it be useful? Bridge maintenance work is meant to restore a bridge

Highway Bridge Maintenance Planning and Scheduling
http://dx.doi.org/10.1016/B978-0-12-802069-2.00006-4

to an improved condition. Typically, this is only economical when deficiencies are limited and contained to particular elements. A bridge which is severely damaged throughout by an event such as a fire or an earthquake, or a bridge that is deficient for current traffic loads throughout all the elements of its superstructure and substructure, may be more suited for replacement than for repair. A bridge also might have a critical element that had been previously repaired and is no longer suited for further repair. An example of this would be a reinforced concrete deck girder, which had been significantly patched several times before. If further repair is no longer possible, replacement is the only option if a bridge is to remain in service at the location.

This may lead to the question, is the bridge necessary? Over the past several decades many rural areas of the United States have experienced a decrease in population density with the consolidation of farming operations. In the same time period, the distribution of both population and industrial activity has changed significantly in many urban areas. Therefore, the traffic demands of a highway served by a bridge may have significantly decreased since its construction. With any drastic reduction in demand, one must consider whether a bridge is necessary before committing funds and effort to repair it. Once repaired, even a bridge in good condition will incur ongoing preventative maintenance costs. If there are alternate routes, which provide access to properties served by the bridge and if traffic volumes on it are minimal, the prudent course may be to remove a bridge from service rather than repair it.

If the bridge is in a rural area, an alternative to completely closing a route would be to accept a lower, more appropriate, level of service. A deteriorated span structure might be replaced with a low water crossing in locations where access to residences or businesses would not be affected.

Two useful pieces of information to help weigh this decision are available on the bridge inspection report, the average daily traffic (ADT), and the bypass detour length. The bypass detour length is National Bridge Inventory (NBI) item 19. It is a report of the adverse travel that would be incurred if the bridge considered were closed.

Third, what is the minimum condition required for operation of the bridge? This requires defining which users must be accommodated with this bridge. Is this a location with little truck traffic where a load posting will have little impact? Or is the bridge, instead, located on a major freight route used by a number of trucks with overweight permits? Aside from the need for load capacity for truck traffic, a sufficient roadway width may be needed to provide operational capacity for traffic volumes. A high degree of deck condition may need to be maintained on routes with heavy volume of traffic to minimize delay times. Is this a necessary route for emergency services? If a bridge cannot facilitate required operations for its users due to damage or deficiency, it must be repaired or rehabilitated.

If a location has a particularly high ADT or high volume of truck traffic, engineering judgment should be used to consider if the typical rates of deterioration for particular bridge elements at this location are reasonable assumptions. If an accelerated rate of deterioration might be expected, this would serve as an impetus to scheduling maintenance work as soon as possible.

Lastly, what is the likely remainder of service life? If a bridge is on a highway which is scheduled to be reconstructed on a new alignment within a few years, it would not be prudent to spend money on repairs unless absolutely required to facilitate traffic in the intervening years. A bridge that is functionally obsolete, or would be so under traffic volumes predicted for future years, may be scheduled for replacement on that criterion.

When making a determination as to the likelihood that a bridge might be replaced before it is required to be for structural reason, planned future use of properties served by the bridge should be considered. Metropolitan Planning Organizations (MPO) and state Department of Transport (DOT) s are required by the federal government in the United States to maintain a transportation improvement plan (TIP) with a minimum of a 4-year window, and a long-range transportation plan (LRTP) with a minimum 20-year window. The TIP and the LRTP, and other planning documents by MPOs, contain projections on future land use. Development of an area might be expected to significantly increase traffic volume at a site.

6.1.2 Adjusting Scope

If the first four criteria indicate that a bridge is a likely candidate for bridge maintenance work, then additional criteria should be considered in determining the scope and the timing of repair work.

1. What is the desired condition for this bridge?
2. Is there a need for similar bridge work in the area?
3. Where does this bridge weigh against others on the basis of need?

There can be considerable difference between the minimal acceptable condition for a bridge and the desired condition. The level of desired condition may not only depend on the range of users that it is desired to serve (such as heavy trucks) but on potential savings in future maintenance costs from maintaining a higher condition level. This presumes that the rate of deterioration decreases with increases in element condition. BLCCA is a technique that can take into account changes in deterioration rates resulting from maintenance action, and the timing and scope of those actions, and predict the lowest-cost scenario of maintenance action timing and scope to maintain a chosen level of condition. By comparing analyses made to maintain various levels of condition, one may find a level that may be maintained at the lowest life cycle cost. Alternatively, one might also determine the life cycle cost difference for accepting a lower level of condition (Figure 6.1).

Costs for contract maintenance work can also be decreased by getting lower prices for the same work. Substantial maintenance work entails significant cost for contractors to mobilize and to provide traffic control at a work site. If work on several bridges on a highway facility or on nearby facilities can be combined into a single project, these costs can be minimized. Also, larger quantities of work may attract more bidders and, consequently, more competitive bids.

Most bridge owners find that the needs for bridge maintenance outweigh the funding available. Although a particular bridge may need maintenance work and the lower costs scenarios of maintenance actions may have been identified, it must compete against the needs of others in its bridge inventory. A system-wide review of bridge maintenance needs is the province of bridge management and will be discussed in the Chapter 7.

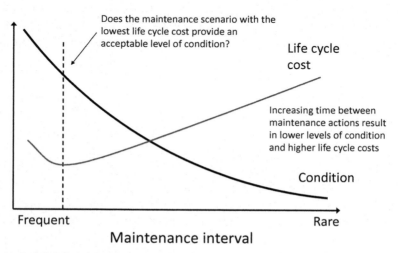

Figure 6.1 Costs and Condition as a Function of Maintenance Effort. *(Adapted from Paul D. Thompson).*

6.2 BRIDGE LIFE CYCLE COST ANALYSIS

Bridge components deteriorate over time. The range of maintenance efforts, from preventative maintenance to rehabilitation, are expended to counter the effects of deterioration and to restore damaged components to an acceptable condition. These maintenance efforts vary in both cost and effectiveness. The more effective efforts slow the rate of deterioration after application. For a typical bridge, a number of different scenarios of maintenance actions can be contemplated over its service life, each representing a different combination of funds spent at varying times throughout that life. Each of these scenarios also differently impacts the rate of deterioration of bridge components resulting in changes in expected service life between scenarios. Comparing each of these alternate scenarios of maintenance actions requires dealing with the value of money as it varies through time.

The FHWA *Life-Cycle Cost Analysis Primer* lists five steps to conducting a life cycle cost analysis [1]:

1. Establish design alternatives;
2. Determine activity timing;
3. Estimate costs (agency and user);
4. Compute life cycle costs; and
5. Analyze the results.

With the possible exception of emergency repair work to heavily traveled bridges, for any bridge at any time there are a number of possible scenarios for maintaining the bridge over time. The owner may choose to do little in the way of preventative maintenance and to allow deterioration to continue unabated until the structure must be restricted or even closed. Or the owner may need a very high level of functionality at a particular location and invest a large amount or resources into almost constant maintenance. By contemplating scenarios of particular maintenance actions at particular times (when will depend on the deterioration rate of the bridge, or of the element under consideration) and estimating the costs for such work, BLCCA can be used to calculate the cost incurred over the length of the analysis period.

As discussed later, the value of money changes through time. The focus of a BLCCA is to convert these expenditures through time into a common base-year value of money. This allows for a fair comparison of alternatives.

6.2.1 Inflation

In the United States, for the past several decades, the prices of goods and services has tended to rise each year, so that it is commonly expected that

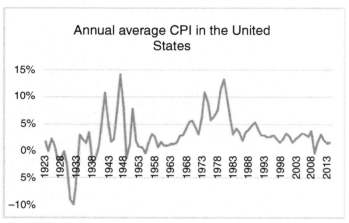

Figure 6.2 *Annual Average Consumer Price Index in the United States from 1923 to 2014. (US Bureau of Labor).*

the purchasing power of a dollar will be less next year than it is now. This reduction in the purchasing power of individual units of money is termed inflation. When prices, in general, fall and purchasing power increases, it is called deflation. According to the data on the consumer price index[1] from the United States Bureau of Labor Statistics, as of 2015, 2009 was the only year with deflation (at −0.4%) since 1955 [1] (Figure 6.2).

Due to this change in the purchasing power of money over time, a fair comparison of the expenditure of funds in different years requires compensating for the effects of inflation by converting the value of dollars in terms of purchasing power from the year spent to an arbitrary base year. For a BLCCA, the base year considered is typically the year the analysis is conducted. When estimating the costs of maintenance actions in the future (typically based on the cost of similar work done this year), it is a simple matter of ignoring the effect of inflation on future cost and leaving the estimate in base-year dollars. If examining past costs for the purposes of estimating, the dollars must be converted to the base year. If a constant rate of price inflation (pi) were assumed, then base-year dollars could be calculated as:

$$\text{Dollars}_{\text{Base Year}} = \text{Dollars}_{\text{Year Spent}} \times (1 + \text{pi})^n$$

---

[1] The consumer price index is a measure of the changes in the cost of living in the United States over time. It is the average month-to-month change in the prices consumers pay for a typical "market basket" of goods and services.

Where n = the number of years between the base year and the year the funds were spent.

However, as noted from the equation above, inflation has not been steady over the years. And, while measures such as the consumer price index are a measure of the purchasing power of dollars in the economy as a whole; for estimating costs for highway bridge maintenance, it is the trends of those particular costs that are of interest.

Since 1922, the FHWA has complied data from contracts awarded to construct federal aid highway projects. From this, they recorded variations in bid prices by quarter in a quarterly published bid-price index (BPI) (Figure 6.3). Although traditionally this has been used for recognizing trends in construction costs and for making adjustments to contracts after award, it can be used as a construction cost index to adjust prices paid in the past year contracts to a base year for comparison by the equation:

$$\text{Dollars}_{\text{Base Year}} = \text{Dollars}_{\text{Year Spent}} \times \frac{\text{Construction Cost Index}_{\text{Base Year}}}{\text{Construction Cost Index}_{\text{Year Spent}}}$$

The annual average BPI is shown for years 1972–2005 later.

To convert the cost of a maintenance action let for contract in 1972–2004 base-year dollars, the 2004 BPI composite index of 154.4 and the 1972 BPI composite index of 38.6 result in a ratio of 154.4/38.6 = 4.0. A maintenance project that let for $10,000 in 1972 would be expected to have let for $40,000 in 2004. Using this ratio of base year to current year dollars over

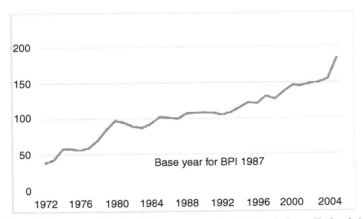

Figure 6.3 FHWA Average Annual Composite Bid-Price Index. *(Federal Highway Administration).*

a 32 year period (1972–2004) in the equation, assuming a constant rate of inflation, results in an equivalent annual inflation of 4.42% over the period.

In 2006, the FHWA updated its construction cost index to a new measure, the National Highway Construction Cost Index (NHCCI) (Figure 6.4). The NHCCI is calculated using a larger database than was the BPI, including a significant amount of information from state high-way projects. The methodology of weighing the selection of bid items used in calculation has also been updated as has the vetting of data used. The data used for calculation of the NHCCI starts in 2003, so there is some overlap between it and the BPI [2].

Note, that though the trend of the NHCCI is generally upwards over time, there was a spike in prices from 2006 to 2010. Using construction contracts awarded in those years as a basis for the estimate of future work will require careful consideration before deflating those prices to a base in a later year.

Most states specify inflation rates to be assumed in estimates for their work, based on their local economic data. Rates from the Kansas Department of Transportation (KDOT)'s website (http://ksdot.org/Assets/wwwksdotorg/bureaus/burStructGeotech/pdfs/inflation.pdf) are shown in Figure 6.5 as an example.

6.2.2 Discount Rate

Separate from the issue of inflation is the consideration of the opportunity value of time. When contemplating future expenditures, it is commonly

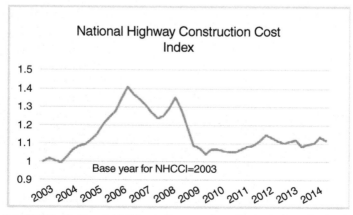

Figure 6.4 FHWA National Highway Construction Costs Index. *(Federal Highway Administration).*

Revised Inflation Rates for Construction Costs - May 2010

From FY	2000	2001	2002	2003	2004	2005	2006	2007	2008	2009	2010	2011	2012	2013	2014	2015	2016	2017	2018	2019	2020
1999	3.8%	7.6%	11.4%	15.1%	19.0%	23.2%	27.7%	32.4%	37.5%	42.9%	48.4%	53.6%	59.0%	64.5%	70.3%	76.3%	84.2%	92.6%	101.1%	110.2%	119.7%
2000		3.7%	7.5%	10.9%	14.6%	18.7%	23.0%	27.6%	32.4%	37.6%	43.0%	48.0%	53.2%	58.5%	64.1%	69.8%	77.5%	85.4%	93.8%	102.5%	111.6%
2001			3.5%	6.9%	10.8%	17.4%	18.7%	23.0%	27.7%	32.7%	37.0%	42.7%	47.7%	52.9%	58.2%	63.8%	71.1%	78.8%	86.9%	95.3%	104.1%
2002				3.3%	6.8%	10.6%	14.6%	18.9%	23.4%	28.2%	33.2%	37.9%	42.7%	47.7%	52.9%	58.2%	65.3%	72.8%	80.6%	88.7%	97.2%
2003					3.4%	7.0%	11.0%	15.1%	19.5%	24.1%	29.0%	33.5%	38.1%	43.0%	48.0%	53.2%	60.1%	67.3%	74.8%	82.9%	90.9%
2004						3.5%	7.3%	11.3%	15.5%	20.0%	24.7%	29.1%	33.6%	38.3%	43.1%	48.1%	54.8%	61.8%	69.0%	76.6%	84.6%
2005							3.7%	7.5%	11.6%	16.0%	20.5%	24.7%	29.1%	33.6%	38.3%	43.1%	49.8%	56.3%	63.3%	70.7%	78.3%
2006								3.7%	7.6%	11.9%	16.2%	20.3%	24.6%	29.8%	33.3%	38.0%	44.2%	50.7%	57.5%	64.8%	72.0%
2007									3.8%	7.8%	12.1%	16.0%	20.0%	24.2%	28.6%	33.1%	39.1%	45.3%	51.9%	58.7%	65.8%
2008										3.9%	8.0%	11.7%	15.6%	19.7%	23.9%	28.2%	34.0%	40.0%	46.3%	52.9%	59.8%
2009											3.9%	7.6%	11.3%	15.2%	19.2%	23.4%	29.0%	34.8%	40.8%	47.2%	53.8%
2010												3.5%	7.1%	10.9%	14.8%	18.8%	24.1%	29.7%	35.5%	41.6%	48.0%
2011													3.5%	7.1%	10.9%	14.8%	19.9%	25.3%	31.0%	36.9%	43.0%
2012														3.5%	7.1%	10.9%	15.9%	21.1%	26.5%	32.2%	38.2%
2013															3.5%	7.1%	11.9%	17.0%	22.2%	27.7%	33.5%
2014																3.5%	8.2%	13.0%	18.1%	23.4%	29.0%
2015																	4.5%	9.2%	14.1%	19.3%	24.6%
2016																		4.5%	9.2%	14.1%	19.3%
2017																			4.5%	9.2%	14.1%
2018																				4.5%	9.2%
2019																					4.5%

Figure 6.5 *Inflation Rates for KDOT.*

assumed that uncommitted funds can be used in a productive manner to earn a rate of return greater than the inflation rate. Such a rate of return might be derived from investing in other capital projects, which are assumed to result in economic benefits, or from investment in financial instruments such as stocks or bonds. The discount rate is this assumed return, less the rate of inflation.

The selection of a discount rate has a considerable impact on the result of the BLCCA. Low discount rates assume a low opportunity cost for money spent today and favor current expenditures, while high discount rates assume that better returns might be found in other investments and reduce the present value of future costs. The discount rate generally selected for the life cycle cost analysis of public transportation projects assumes the conservative rate of return of secure long-term bonds such as the 30 year United States Treasury Notes and Bonds, less the projected rate of inflation.

Projects by agencies of the United States government follow the guidelines set forth by the Office of Management and Budget in *Appendix C of Circular A-94*, which bases the discount rate on the real interest rate for United States Treasury Notes and Bonds. In 2015, the real interest rate specified in the Appendix for the 30 year note was 1.4%. However, the standard practice for BLCCA, as outlined in NCHRP 12-43 is to use a discount rate between 2–4% [3]. This recognizes historical trends in the value of money.

To assess the sensitivity of BLCCA to the discount rate assumed, it is common to conduct the analysis for two or three likely rates and compare the results.

6.2.3 Cash Flow Diagrams

With anticipated expenditures converted to base-year dollars, they may be plotted on a timeline to produce a cash flow diagram (Figure 6.6).

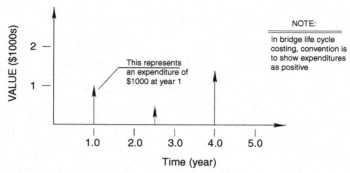

Figure 6.6 *Sample Cash Flow Diagram.* Note: In bridge life cycle costing, convention is to show expenditures as positive.

Major expenditures, such as initial construction, repairs, and rehabilitation, are shown as occurring at a point in time. Ongoing expenses, such as preventative maintenance are shown as distributed costs over a period of time. The diagram allows an easy comparison of costs by use of the visual scale.

As seen in the discussion in Chapter 4, the condition of a particular bridge component, or of the overall bridge may also be plotted along a timeline. The service life is the time from the initial construction until a failed condition state is reached. Without any maintenance interventions, the condition diagram is similar to the one shown in Figure 6.7.

Combining the cash flow diagram and the condition diagram on the same time scale allows the visual representation of the life cycle activity of the bridge or bridge component under consideration as seen in Figure 6.8.

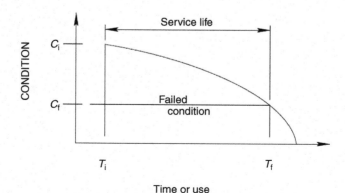

Figure 6.7 *Simple Condition Diagram.*

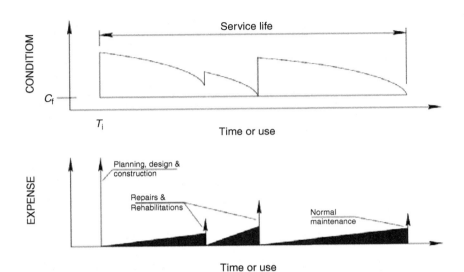

Figure 6.8 *Example Life Cycle Cost and Activity Diagram.*

The time period analyzed by BLCCA for maintenance actions typically does not last for the expected service life of the structure. Typically, a highway bridge has an expected life of decades. Before the widespread use of epoxy-coating reinforcing steel and other protective methods in bridge construction, a 50-year service life was commonly assumed. Since then, a 75-year service life is expected for most bridges. The latest AASHTO specifications call for a 100 year design. NBI records have been mandated only since the early 1970s in the United States, so predictions of deterioration and the cost of repair work are based on observations over a period of time less than the service life of bridges based solely on deterioration rates. Also, differing maintenance scenarios result in different expected service lives.

Service life may be cut short before replacement is mandated by failed condition by obsolescence. A bridge may be replaced due to insufficiency in characteristics such as roadway width, which impair the traffic capacity and safety of the bridge. NCHRP 12-43 recommends the selection of a review period that includes at least one projected major repair or rehabilitation project. As a practical matter this results in a range of review periods extending from 20–40 years.

6.2.4 Residual Value

Given that a bridge will typically be in service at the end of the BLCCA review period, the bridge in place will have some remaining value that must be accounted for in the analysis. This value is often, incorrectly, referred to as the salvage value. If a bridge is to be removed at the end of a review period, then the value of material sold for reuse or recycle, less the removal cost, is the actual salvage value of the bridge. Commonly though, the removal cost of a highway bridge exceeds any revenue to be had from recycling of material. To calculate a residual value to be used in BLCCA, the FHWA recommends proportioning the cost of the last major rehabilitation work (or initial construction cost if no rehabilitation has been done) by the remaining expected service life at the end of the analysis period.

6.3 DETERMINING COSTS

When determining which costs to include in a BLCCA it must be remembered that the analysis is meant to compare the economic efficiencies between alternate scenarios over a period of time. It is not meant to be a tool for budgeting. Only the costs which differ between scenarios need be considered. Costs incurred over the life cycle, which are the same may be excluded from the analysis. This includes sunk costs, that is, monies which have already been expended by the beginning of the analysis period considered.

6.3.1 Agency Costs

Three types of costs are considered in BLCCA: agency costs, user costs, and vulnerability costs.

Agency costs are those paid by the bridge owner to build, maintain, or remove the bridge. These costs include contract payments from the owner other entities (e.g., construction and consulting contracts) and operational and administrative costs. Preliminary engineering, contract administration, and preliminary maintenance activities are all agency cost associated with bridge maintenance work. Since these costs are budgeted by the owner, there are generally existing good records on which to base estimates of future work, particularly so for those costs paid on contracts.

Because these are the costs which impact bridge owners' budgets, they are included in all BLCCA and are often the only costs considered. Although user costs may be a significant consideration on some projects, the need to maximize the effectiveness of a limited agency budget is a consideration on every project.

The typical agency costs that included in BLCCA between life cycle alternatives are those associated with either construction projects (for substantial maintenance work, replacement, or removal) or for maintenance (typically preventative maintenance by in-house personnel). Operational costs such as bridge inspections are common to all scenarios where either the existing, or a new, bridge is maintained in place and are not included in cost comparisons.

Construction costs include the engineering efforts to prepare plans, the costs paid on the let contract and the costs to administer the contract and observe the construction. The construction costs of proposed projects can be estimated by assuming quantities of individual bid items that will be required for the work and applying unit costs for each items based on the average bid price to date.

For example, consider a simple project involving the deck patching and overlay of a bridge deck with a thin polymer concrete overlay. The significant typical bid items involved would be:
- Area prepared for patching (square yards);
- Area prepared for patching – full depth (square yards);
- Polymer concrete overlay (square yards);
- Mobilization (lump sum); and
- Traffic control (lump sum).

Mobilization is payment for the contractor's costs to accept the job and bring the necessary equipment and prepare the site for the project. Traffic control refers to the activities necessary to either maintain or divert traffic at the project site. Different bridge owners may break out and track these costs with differing combinations of specific bid items.

A deck NBI condition rating is 7 indicates that less than 10% of the deck surface is deteriorated. If a maintenance project is to be scheduled at the cusp of deterioration from a rating of 7–6, it can be presumed that 10% of the deck will require patching. For a girder bridge, perhaps 10% to this area will be full depth patching. For a girder bridge that is 120 ft. long with a 36 ft. roadway width an estimated project construction cost is shown in Table 6.1.

Engineering and administration costs are commonly estimated as a percentage of the construction costs, typically 15–20% for bridge maintenance projects. To calculate life cycle costs for a maintenance strategy that includes applying polymer overlay when the deck condition reaches 10% deterioration assume a project cost of approximately $104,000 applied in the desired year.

Note, to allow for quick estimating, it is common to track awarded contract bids for bridge work by square ft. of bridge deck. For a project

Table 6.1 Example estimate of bridge deck patching and polymer overlay

Bid item	Unit	Unit cost ($)	Quantity	Price ($)
Area prepared for patching	Square yards	500	43	21,500
Area prepared for patching- full depth	Square yards	700	5	3,500
Polymer concrete overlay	Square yards	70	480	33,600
Mobilization	Lump sum	15,000	1	15,000
Traffic control	Lump sum	15,000	1	15,000
			Total	88,600

such as this, an estimate of $20–$25 per square ft. (in 2015 dollars) might be expected.

Costs for maintenance activities by the owner's in-house crews can be estimated from crew timesheets and equipment usage logs, if such activity is tracked by individual structures. Often this is not the case, and even when such tracking is the standard protocol compliance by crews may be inconsistent. A better course is to identify the common in-house maintenance actions calculate standard costs based on the resources used and the usual speed of work. Examples of such calculations for deck sealing and deck patching by KDOT forces in 2010 are shown in Tables 6.2–6.4.

Converting to a square ft. cost, sealing a concrete deck may be estimated at $825 per 8000 of bridge deck = $0.10 per square ft. Patching a bridge deck with cold mixed asphalt will be approximately $40 per square ft. of patching required and patching with rapid set concrete will be approximately $80 per square ft.

Table 6.2 KDOT agency cost to seal typical 8000 sq. ft. deck per day

Equipment	Quantity	Rate ($)	Cost per day ($)
Pickup truck	1	45	45
Deck sealer	1	172	172
Material			
High molecular weight methacrylate sealer			200
Labor			
2-Equipment operators	2	204	408
		Total per day	825

Table 6.3 KDOT agency cost to bituminous patch 45 sq. ft. of bridge deck per day

Equipment	Quantity	Rate ($)	Cost per day ($)
Pickup truck	1	45	45
Dump truck	1	190	190
Cold mix equipment	1	318	318
Material			
Cold mixed asphalt	45 sq. ft.	6	270
Labor			
Supervisor	1	252	252
3-Equipment operators	3	204	612
		Total per day	1687

These costs are contingent on agency wage rates, equipment availability, and material costs. The equipment rate for the vehicles above assumed roundtrip distances of 30 miles to the job site. This may be a significantly unconservative assumption in rural areas. Traffic control measures, which are dependent on the Average Annual Daily Traffic (AADT) of a particular location, are not included in the above estimates. The maintenance costs above are shown as an example of what may be calculated for a particular geographic area for a particular agency.

When calculating maintenance activity costs for BLCCA, it must also be noted as to what is the expected life of the action. Deck sealing may be applied every 1–2 years. Deck patching by in-house maintenance personnel,

Table 6.4 KDOT agency cost to concrete patch 27 sq. ft. of bridge deck per day

Equipment	Quantity	Rate ($)	Cost per day ($)
Pickup truck	2	45	90
Dump truck	1	190	190
Hammers and vibrators	1	292	292
Material			
Rapid set concrete	27 sq. ft.	9	243
Labor			
Supervisor	1	252	252
5-Equipment operators	5	204	1020
		Total per day	2087

particularly asphalt patching, is not expected to last as long as patching done under contract for substantial maintenance actions. In an urban area at a site with a large AADT, asphalt patches might last only a few months. In a rural area with low AADT, concrete patching may last a few years. The life of maintenance work can be highly variable by agency and location and should be determined by local experience.

The amount of documentation and data available to estimate agency costs may lead to overconfidence in the ability to project the cost of future activities. In the literature and research on BLCCA, the ability to develop maintenance work scenarios 40 years in the future is not questioned. Although, the rate of inflation for highway construction has not differed greatly from the rate of inflation in the United States as a whole over the past few decades, in the same time period new products and constructions techniques have entered the practice of bridge construction. New types of deck overlays, expansion joints, bearings, reinforcing for concrete, and other materials promise to reduce deterioration rates and extend bridge service life. Additionally, new costs associated with construction, such as requirements for environment mitigation have become part of standard practice.

Although the practice of engineering, constructing, and maintaining highway bridges tends to be conservative, with changes in practice taking years and decades to become widespread, these changes do and will continue to occur. One should be cautious in projecting agency costs for several decades in the future based on historic costs.

6.3.2 User Costs

Most common bridge maintenance activities involve work at the roadway, resulting in disruption of the normal flow of traffic. Traffic must be diverted at the work zone, either onto adjacent lanes or on designated detour routes around the site. This imposes direct costs on the intended user of the bridge in the form of increased travel time due to delay and, for detoured traffic, in form of extra distance that must be driven (also known as adverse travel).

The lowest agency cost for bridge construction work is usually achieved when the bridge is closed to traffic. Facilitating traffic through a construction area on a bridge requires traffic control measures of either carrying traffic at site on an adjacent temporary structure, or of phasing the construction work and carrying traffic on adjacent open lanes. For a bridge with two or more lanes traveling the same direction, one lane can be closed and the entire volume of traffic on the open lanes. For a two-lane bridge carrying traffic each direction, traffic may be phased with signalization. Either of these phasing options require traffic control measures, prolong the length of time needed for construction and require the contractor to remobilize for each phase of construction. This can be considerably more expensive than closing the bridge to traffic during construction. However, closing a bridge can impose a significant inconvenience on the travel public. Monetizing this inconvenience in BLCCA, calculations can provide a means of assessing the impact to users and comparing to it to the budget impact on the bridge owner. This may be used to justify either maintaining traffic through construction or accelerating the construction work.

Of the three possibilities for handling traffic at a work zone, the easiest to calculate user cost for is the detour option. First the adverse mileage is calculated. This is simply the length of the detour less the length of the normally traveled route. The additional travel time is also calculated. This may be as simple as finding the difference between traveling the detour route at its signed speed and traveling the normal route at its usual signed speed. Or, the calculation may be expanded to include delay time is slowing and stopping. The FHWA's *Work Zone Road Users Costs- Concepts and Applications* report is the most recent national guide to analyzing user costs in work zones and contains instructions on calculating vehicle operation costs (VOC) in terms of per mile and per hour costs [4]. Many state departments of transportation will provide standard road user cost for passenger vehicular and trucks for use by consulting engineers. The cost imposed on each user is the length of adverse travel times the VOC per mile plus the time of adverse travel times VOC per hour.

Calculations for VOC per mile include consideration of fuel and oil usage and depreciation of the vehicle. Calculations for VOC per hour include consideration of the value of time for the occupants and of any freight being carried. The reader is referred to the FHWA guide for specifics in calculations. In a study of the KDOT substantial maintenance practices in 2013, VOC for passenger vehicles of $17.25 per hour and $0.62 per mile and VOC for trucks of $31.95 per hour and $1.18 per mile were calculated [5].

The alternative to detouring traffic is to carry it through the construction site. If the site has more than one lane in each direction, the simplest phasing scheme is to reduce the number of lanes for traffic. In this case, determining the user cost requires calculating how many vehicles are expected to be delayed due the traffic demand exceeding the reduced capacity through the bridge site, and what the average delay is expected to be. For urban areas, peak hour traffic volumes may be available for major routes from the local highway agency. If not, the peak hour volume may be estimated by multiplying the AADT by an hourly traffic variation factor. In the referenced study of KDOT, the peak hour factor reported by the agency was slightly less than 9% [5].

The capacity of the lanes carrying traffic through construction may be determined by the methodology found in the section, *Capacity Reductions due to Construction and Major Maintenance Operations* of the 2010 *Highway Capacity Manual* (HCM) [6]. Exhibit 10-14 of the 2010 HCM provides values for lane capacities of long-term construction zones. For a reduction of two lanes to one, the single lane has a default capacity of 1400 vehicles[2] per hour. For reductions of three lanes to two, each remaining lane has a default capacity of 1450 vehicles per hour.

There are three adjustments provided for the base-lane capacity values. The first is for the effect of heavy vehicles in the traffic stream. A heavy-vehicle adjustment factor is provided that is a function of the proportion of trucks and of recreational vehicles in the traffic stream. A second adjustment is for the presence of ramps. The third adjustment is for lane widths. In lieu of reducing the capacity of the lane considered, the traffic volume may be increased by counting trucks as 1.5 passenger vehicles. To determine if a queue will form, resulting delay at a work zone, compare the capacity of the lane or lanes to the passenger vehicle equivalent traffic.

As an example, consider a deck repair project to a pair of two-lane highway bridges, each carrying directional traffic with an AADT of 14,100 vehicles and 5% truck traffic. If there are no ramps in the work zone and 12 ft. lanes are maintained, the hourly capacity of a single lane in a work zone is 1400. The peak hour volume may be estimated as $0.09 \times 14{,}100 = 1269$. Converting the truck traffic to equivalent passenger cars ($1269 \times 0.05 \times 1.5 = 95$) results in an equivalent hourly traffic volume of $0.95 \times 1269 + 95 = 1301$. The volume to capacity ratio is $1301/1400 = 0.93$, which is less than 1.0, so no delay is expected.

If the peak hour traffic volume does exceed the work zone capacity a closer look must be taken at the hourly distribution of traffic to determine how many hours a day volume exceeds capacity. Delay can be calculated by assuming that vehicle arriving at the work zone in excess of capacity are queued and then discharged at the lane capacity as arrivals taper. Table 6.5 shows the results for the calculation of a daily user cost from queue lengths and delays for a location with an AADT of 16,000 and 5% truck volume based on KDOT hourly traffic variation factors for weekday traffic. Using a methodology based on queueing theory as discussed by Jiang the maximum delay time in any hour is the number of queued vehicles present in that hour divided by the discharge rate [7].

---

[2] In this usage, vehicles refer to passenger automobiles rather than trucks.

Table 6.5 Sample delay calculation at work zone

Total number of lanes	2	Number of lanes open	1
Truck percentage	5	Open lane capacity	1600
Total capacity of work zone	1400		
Hourly user cost			
Semi-trailer trucks	$31.95		
Passenger cars	$17.25		

Time	Demand	Capacity	Total arrivals	Total departures	Queued vehicles	Queue length	Delay (min)
0:00	123	1,400	123	123	0	0.00	0.00
1:00	72	1,400	195	195	0	0.00	0.00
2:00	62	1,400	257	257	0	0.00	0.00
3:00	62	1,400	320	320	0	0.00	0.00
4:00	136	1,400	456	456	0	0.00	0.00
5:00	351	1,400	807	807	0	0.00	0.00
6:00	817	1,400	1,624	1,624	0	0.00	0.00
7:00	1,348	1,400	2,972	2,972	0	0.00	0.00
8:00	1,012	1,400	3,984	3,984	0	0.00	0.00
9:00	772	1,400	4,756	4,756	0	0.00	0.00
10:00	754	1,400	5,510	5,510	0	0.00	0.00
11:00	840	1,400	6,350	6,350	0	0.00	0.00
12:00	894	1,400	7,244	7,244	0	0.00	0.00
13:00	907	1,400	8,151	8,151	0	0.00	0.00
14:00	1,010	1,400	9,161	9,161	0	0.00	0.00
15:00	1,238	1,400	10,399	10,399	0	0.00	0.00
16:00	1,415	1,400	11,815	11,799	15	0.04	0.66
17:00	1,473	1,400	13,287	13,199	88	0.22	3.77
18:00	941	1,400	14,229	14,229	0	0.00	0.00
19:00	635	1,400	14,863	14,863	0	0.00	0.00
20:00	530	1,400	15,393	15,393	0	0.00	0.00
21:00	459	1,400	15,852	15,852	0	0.00	0.00
22:00	330	1,400	16,182	16,182	0	0.00	0.00
23:00	221	1,400	16,403	16,403	0	0.00	0.00

3.77	Max delay (min)
0.38	Average delay (min)
$102.59	User cost
0.22	Max queue length (miles)

Note: Hourly traffic volumes above are adjusted for equivalent trucks.

In this example, the peak hour volume only slightly exceeds the work zone lane capacity. The maximum delay during rush hour is only less than 4 min. The calculated user cost is approximately $100 per day given the VOC assumed for Kansas. Calculating the user costs with the same VOC for a 0.5 mile detour, instead, results in a user costs of approximately $5400 per day. This example is not atypical. Phased construction may significantly extend the working days required to complete a project, but it has considerably less user-cost impact than even short detours.

For signalized work zones, the delay calculation become more complex. An example of calculations for delay and user cost for a signalized work zone on a 300 ft. long bridge in relatively low traffic area is given in Appendix 1.

The calculations for delay shown earlier and in the appendix are truly rough estimates. To better calculate work zone delays that might be expected, the involvement of an experienced traffic engineer and a site assessment would be required for each project. Complex bridge sites might require microsimulation modeling of local traffic patterns to determine the likely responses of local drivers to work zones. However, the examples serve to illustrate the scale of user impacts. In urban areas with higher traffic volumes, the user costs of detouring traffic can readily compare to, or exceed, the agency costs of substantial maintenance work for bridges. As a result, in these areas, phased construction is standard practice for most bridges with high traffic volumes where possible.

Although NCHRP Report 483 recommends including the increased risk of crashes in work zones in the user-cost calculations it is difficult to so with reliable data [9]. Acquiring such data would require quantifying the increase in crash rates that might be expected from the presence of a work zone. Bridge owners, which must employ work zones on a frequent basis, are loathe to do so because of the liability implications in future litigation. As stated by Bai in a report on causes of work zone crashes for KDOT, "Work zone crash rates by work zone travel mileage are not precisely known" [8].

An item to note from Bai's report is that only 5% of work zone crashes occurred at bridge sites. Bridge projects tend to be fairly short in comparison to other maintenance projects involving pavement rehabilitation or shoulder reconstruction. Travelers are exposed for a relatively limited time at sites involving bridge construction.

However, the benefit in the long-term reduction of crash rates due to substantial maintenance work on bridges has been quantified. In a report for the Florida DOT, Gan and Shen researched crash reduction factors calculated and used by state DOTs across the United States [9]. They report that three states, Indiana, Kentucky, and Missouri, had calculated crash reduction factors for the repair of bridge decks. Each found crash reductions (no distinctions made between fatal, injury, or property damage only crashes) of 13, 14, and 15%, respectively. This work was also cited by FHWA in their 2008 reference work for crash reduction factors [10].

6.3.3 Vulnerability Costs

NCHRP Report 483 also recommends including vulnerability cost. These are the costs associated with the risks of rare and extreme events such as earthquakes or fire. Because of the low risk of such events in any given year for any particular structure, and the typically large costs to mitigate such risks, bridge owners typically deal with these risks as a matter of policy than by assessing the cost efficiency of maintenance strategies. For example, in a state with a risk of earthquakes, high-value routes are selected and structures on those routes are hardened against seismic motions.

6.3.4 Economic Costs

A BLCCA may also take into account the economic impact of the alternatives considered on commercial activities affected by any traffic associated with the bridge. There may be specific cases, such as where access to a business would be closed by closure of a bridge, but in general the economic impacts of deteriorated bridge condition cannot be fully captured in life cycle cost analysis. An economic impact analysis is required which takes into account changes in traffic pattern, and the economic value of that traffic. An example analysis was conducted as part of the 2013 study of substantial maintenance practice in Kansas and is included in Appendix 2. The analysis consisted of studying the impact of deferring maintenance on a pair of bridges carrying K-10 near Desoto, Kansas to the point that the bridges would be restricted from allowing the passage of extralegal truck loads on that segment of K-10. In the analysis, the passage of legal loads was considered to be unaffected. The user costs of the restrictions would have been $1.6 million and the impact to the community in terms of decreased business output and lost wages would have been $488,000 for the 2-year period in the analysis.

6.4 DETERIORATION RATES

The key to projecting when maintenance actions will be required in the future is the ability to predict the future condition of bridge elements. Until several years of consistent NBI inspection data was available in the 1980s, the only method to predict bridge deterioration was expert opinion. With several years of NBI data the development of mathematical deterioration models specific to bridge elements was possible.

With accurate deterioration modeling being a key to effective BLCCA and to implementing bridge management systems for planning, there have been several studies and projects to develop models for various states. One of the most comprehensive of these was the March 2009 report, *Bridge Element Deterioration Rates* for the NYSDOT [11]. New York State has maintained a bridge inspection program for over 17,000 bridges in the state since 1981 utilizing a condition rating system for bridge elements. The condition ratings, shown in Table 6.6 (adapted from the report) look very similar to those utilized by FHWA in the NBI inspections, but are not an exact match.

The elements defined in NYSDOT inspections are much more numerous and specific than the more general five NBI elements of deck, superstructure, substructure, channel (for water crossing), and approach for span structures. NYSDOT elements include joints, bearings, pier caps, pier columns, etc. in addition to the general NBI elements − up to 47 distinct elements of a bridge.

The researchers took the bridge inspection data from more than 25 years and analyzed it with the aim of determining the probability that a given element would drop a condition level from one inspection to the next. There are two approaches to determining bridge element deterioration rates: deterministic and stochastic. Deterministic analysis employs techniques of regression analysis to determine the relation between explanatory variables and the progress of deterioration. This approach, though, ignores the effect of unobserved, and therefore unaccounted for, explanatory variables in the process. It also fails to account for the

Table 6.6 NY State DOT condition ratings for bridge elements

Rating	Description
9	Condition and/or existing unknown
8	Not applicable
7	New condition, no deterioration
6	Used to shade between ratings of 5 and 7
5	Minor deterioration, but functioning as originally designed
4	Used to shade between ratings of 3 and 5
3	Serious deterioration, or not functioning as originally designed
2	Used to shade between ratings of 1 and 3
1	Totally deteriorated, or in failed condition

random nature of demands (i.e., traffic loadings) and environmental factors. Stochastic analysis treats deterioration in condition as subject to one
or more random variables. The authors of the report for NYSDOT model
the deterioration process through both a Markov chain approach and a
Weibull distribution approach [12].

The result of the Markov chain analysis is a transition matrix of
probabilities that an element will deteriorate from one condition level
to the next between inspections. This modeling of discrete transitions
makes Markov chains a natural fit for developing deterioration models
from bridge inspection data. Since the 1990s, most bridge management
software with deterioration modeling capabilities have utilized Markov
chains.

An often-cited weakness of the Markov chains approach, though, is that
the probability of transition from one level to the next is the same regardless of the time that a particular element has been at a condition level. As an
example, a deck rated at a condition level of 6 for four previous inspection
cycles has the same probability of dropping to a 5 as a deck that changed to
a 6 in just the last cycle. This can lead to underestimation of deterioration
rates, particularly at the lower condition ratings.

The Weibull distribution approach models the condition rating as a random variable with the probability that the time an element has been in a
particular condition defined by a survival function. Data are analyzed with
an end of determining the shape and scale parameters in the survival function, which leads to a determination of the failure rate. The failure rate may
increase or decrease with the length of time that an element has been in a
particular condition rate.

For structural decks with uncoated reinforcing, the researchers for the
NYSDOT reported that the time for a drop in condition rating from 7–4
was 49 years according to a Markov chain analysis and 43 years with a
Weibull distribution analysis. For structural decks with epoxy-coated reinforcing, the numbers were 62 and 60 years, respectively. One item to note
in the review of the various deterioration rates is how linear the rates of deterioration are with respect to time, as seen in Figure 6.9, which was Figure
4-21 in the NYSDOT report.

In reviewing the NYSDOT data, it is apparent a condition rating 4
is where significant repairs and construction actions are expected to be
required. Deterioration equations derived using the Weibull distribution
analyses are provided for most elements. The time between NYSDOT condition ratings 7 and 4 for common bridge elements are shown in Table 6.7.

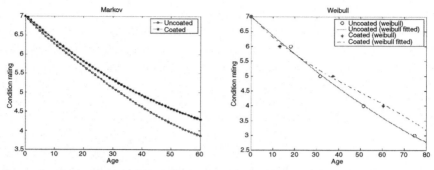

Figure 6.9 *Deterioration Plots for Structural Decks Owned by NYSDOT.*

Note, for concrete box girder bridges and slab bridges, the deck slab is the primary superstructure member. A review of the NYSDOT data shows that the deck (for deck slabs with uncoated reinforcing steel) is the first

Table 6.7 Time to condition Level 4 for bridge elements of NY bridges

	Time for drop in NYSDOT condition rating from 7 to 4	
	Element	**Years**
Bearings	Steel pier bearing	43
	Steel abutment bearing	45
	Elastomeric pier bearing	51
	Elastomeric abutment bearing	59
Joints	Strip seal pier joint	20
	Preformed joint at abutment	24
	Sliding plate pier joint	25
	Modular pier joint	28
	Sliding plate joint at abutment	34
Abutment	Abutment pedestal	55
	Abutment backwall	57
	Abutment stem	58
	Abutment wingwall	58
Pier	Concrete pier cap	55
	Concrete pier pedestal	56
	Concrete pier column	57
	Pier stem	60
	Pier footing	62
Superstructure	Slab or box primary member	47
	T- or I-beam primary member	55
	Rolled beam primary member	56
	Plate girder primary member	60
	Box culvert	56

major structural element liable to need repair or substantial maintenance. Abutments, piers, and girder superstructures all have a life cycle period of 55 or more years before major work is needed, as opposed to 42 year for bridge decks with black steel.

This concurs with general experience that the deck and the deck joints are the first elements to require substantial maintenance on most bridges. These are the elements directly subjected to traffic and to road salts. Any life cycle study focused on a period less than the entire presumed 75 year service life of a bridge will need to pay special attention to deck work.

Other states and researchers have conducted research on deterioration rates of bridge elements, and of decks in particular. Their studies and results regarding the deterioration rates of bridge elements are discussed later in the chapter, in brief.

Development of Agency Maintenance, Repair & Rehabilitation (MR&R) Cost Data for Florida's Bridge Management System, FDOT Contract BB-879, July 2001 – This was the first large-scale review of inspection and construction cost data to include element-level inspection data in the development of MM&R cost data for the Pontis Bridge Management System. A deterioration model of bridge elements was developed by soliciting expert opinion from FDOT personnel experienced in bridge maintenance. The service life of concrete bridge decks was determined to be 50 years. The service life of superstructures was 59 years and substructures 52 years. Joints and bearings had services lives of 21 and 50 years, respectively. Considering that Florida has a fairly aggressive environment for corrosion due to the humidity and the salt spray near the coasts on three sides of the state, the service lives for super- and substructures may represent a lower bound for those elements [13].

Bridge Deck Service Life Prediction and Cost, Virginia Transportation Research Council Report VTRC 08-CR4, December 2007 – Service life estimates of concrete bridge decks and costs for maintaining decks for 100 years were developed. Service life estimates for decks were based on presumed rates of chloride diffusion, which were validated on 10 test decks in the field. The time to reach a deterioration level of 2% for decks with a water/cement ratio of 0.47 was 37 years on average for black steel decks. The time for progression of deterioration from 2–12% (assumed as the end of service life) was 16 years, establishing a typical service life of 53 years. Concrete overlays were found to have lives of 20–26 years and the life was relatively independent of ADT. Polymer concrete overlays lives, however, were highly dependent on ADT, with a range of 10 (ADT over 50,000) to 25 (ADT under 5,000) [14].

Developing Deterioration Models for Nebraska Bridges, Nebraska Department of Roads (NDOR) Project No. SPR-P1(11)M302, July 2011 – Deterioration models for Nebraska bridges were developed from NBI condition ratings from a period of bridge inspections extending from 1998 to 2010. Analysis was performed using Markov chains. Nebraska bridge inspection practice differs from Kansas in that a rating of 9 is assigned to new bridge elements and 5 is the minimal acceptable condition before assigning substantial maintenance. Bridge decks with black rebar were found to have an average service life to condition 5 of 40 years, for those with epoxy-coated reinforcing it was 68 years. ADT was found to not have a pronounced effect except at very low, >100 counts [15].

Steel and prestressed concrete superstructures and concrete substructures were all found to have service lives of 80 years or more. Interestingly, according to the Nebraska paper's authors, the previous approach utilized to predict bridge element deterioration in Nebraska was to drop the deck condition one level every 8 years, and the superstructure and substructure condition levels once every 10 years. An approach it was claimed based on national deterioration rates.

Analysis of Life Cycle Maintenance Strategies for Concrete Bridge Decks, Journal of Bridge Engineering, ASCE, May/June 2004 – Deck substantial maintenance strategies were evaluated using a mechanistic model of concrete deterioration and 8 years of element level inspection data. Concrete overlays were estimated to provide a service life of 15–20 years; bituminous overlays over waterproofing membrane were estimated to provide a service life of 7 years and bituminous overlay without an underlying membrane to provide a service life of 2–3 years [16].

Guidelines for Selection of Bridge Deck Overlays, Sealers and Treatment, part of NCHRP Project 20-07 Task 234, Wiss, Janney, Elstner Associates, May 2009 – Information regarding practice and service life for bridge decks and maintenance treatments was obtained from surveying state DOTs and review of the literature. The surveyed estimate of the service life of low slump concrete overlays had a mean of 16–32 years. For latex-modified concrete overlays the mean of the estimated service life was 14–29 years. Bituminous overlays with an underlying waterproofing membrane had a mean estimated service life of 12–19 years. For a polymer concrete overlay the mean was 9–18 years [17].

Bridge Preservation Guide, FHWA Publication No. FHWA-HIF-11042, August 2011 – This primer on the framework and definitions for bridge-preservation activities provided commonly used frequencies of 10–15 years for applying polymer concrete overlays, 10–15 years for applying bituminous overlays with waterproofing membranes, and 20–25 years for concrete overlays (including silica fume and latex-modified) [18]. These several research results are summarized in Table 6.8.

Although a bridge owner may develop a deterioration model for bridges based on the owner's specific bridge inspection records, there are a number of papers and studies, which provide models. As seen in Table 6.8, most of these are in reasonably close agreement. And, most are in agreement that, at least in the first half of the service life of the bridge elements studied, the rate of deterioration is relatively linear with respect to condition ratings. For the purposes of a BLCCA conducted for a specific project by a bridge owner, it would be a reasonable assumption to take the expected service life for a bridge element and assume a linear rate of deterioration over that life.

Note, that this would not as accurately reflect the true deterioration rate of bridge elements as the models currently employed in bridge management software such as AASHTOWare's BrM; however, the degree to accuracy is sufficient for project maintenance project timing on individual projects.

Table 6.8 Element service lives from the literature

	Service life in years						
Element	Florida	New York	Virgina	Nebraska	Wisconsin	WJSE Survey	FHWA
Concrete deck- black reinforcing	50	49	53	40			
Concrete deck- epoxy reinforcing		60		68			
Concrete overlay		25	22–26		15–20	16–32	20–25
Bituminous overlay with membrane					7	12–19	10–15
Polymer concrete overlay			10–25			9–18	10–15
Superstructure	59	Ave. 57		80			
Substructure	52	Ave. 58		80			

6.5 APPLYING BRIDGE LIFE CYCLE COSTING

6.5.1 BLCCA Programs

There are three stand-alone programs of note to use for BLCCA. They are: the BLCCA program distributed with NCHRP Report 483 *Bridge Life-Cycle Cost Analysis*, BridgeLCC from the National Institute of Technology and Standards, and RealCost from the FHWA. Links to all three are available on the FHWA's website on life cycle cost analysis at: www.fhwa.dot.gov/infrastructure/asstmgmt/lcca.cfm.

Each of these programs was originally released prior to 2004. The only program currently being maintained is FHWA's RealCost. Although this program was originally developed for pavement life cycle cost analysis, it is readily usable for other transportation projects, including those involving bridges. RealCost can calculate agency and user cost and can internally calculate work zone user costs based on FHWA methodology.

One feature of note that is shared by all three programs is the ability to assign a probability distribution to the values of cost used in the analysis. Rather than assign a set value for each cost, the user can assign a range of expected costs and pick a probability distribution (such as uniform, normal, or triangular) based on the user experience with the usual range of bids encountered in contracting. The programs are then capable of running Monte Carlo simulations to determine a range of expected base-year costs.

For BLCCA where there are some significant cost that may be quite variable, or when user costs may make a difference in the alternate selected the use of one of these programs may be useful. However, for most BLCCA conducted by bridge owners, spreadsheet calculations will suffice; particularly when maintenance costs collapse into discrete costs at points in time rather than distributed over a period of time. Spreadsheets offer the advantage of requiring the analyst to explicitly know the assumptions made for the analysis.

BLCCA is meant to compare alternate scenarios by cost, not to provide values for budgeting. User impacts and vulnerability costs for maintenance work are usually dealt with by policy, that is, traffic accommodations through construction are provided in high AADT areas and emergency routes are designated that are to be operable after extreme event. The typical use of BLCCA for maintenance is to determine the most cost-effective scenario for the agency's budget.

As will be discussed in Chapter 7, modern bridge management programs such as AASHTOWare's BrM (formally known as Pontis) include the ability to conduct BLCCA for either a single bridge project or to examine inventory-wide maintenance strategies for a bridge owner. The programs utilize the agency's bridge inspection data on file with the program and deterioration models for bridge element internal to the program, with cost information provided by the agency, to conduct the analysis. As of yet, this functionality is not widely used by bridge owners.

6.5.2 Typical Applications of BLCCA

There are two common uses of BLCCA in the realm of bridge maintenance programming:

1. To determine whether to repair a bridge in the latter half of its expected service life or to replace it.
2. To compare the economic efficiencies of alternate maintenance strategies, such as the use of polymer versus concrete overlays, by conducting a BLCCA on a sample bridge or set of sample bridges.

The first use may be best illustrated by means of a real life example. During contracted deck repairs to a reinforced concrete deck girder bridge carrying K-177 over Munker's Creek near Council Grove, Kansas, the deterioration to the existing parapet and deck at the curb line was found to be significant enough so as to require removal of the parapet with the bridge railing (Figure 6.10). The reinforcing steel had corroded to the point where the parapet was liable to separate from bridge with a vehicular impact. Almost all the reinforcing steel between the deck and the parapet would have

Figure 6.10 *Deteriorated Reinforcing Steel in Parapet (KDOT).*

to be replaced. The most economical repair alternative would be to remove and replace the entire length of bridge rail and parapet on each side of the bridge. This added work would cost more than the bridge work in the original contract bid.

Considering the age of the structure, it was not apparent as to whether replacement of the rail would be a more efficient use of funds rather than installing temporary barrier and replacing the bridge as soon as prudently possible. Paul Kulseth, Bridge Management Engineer for the KDOT conducted a BLCCA comparing two scenarios with a review period ending in 2070:

1. Proceeding with a cost overrun of the current substantial maintenance project and replacing the existing bridge rail and parapet with new permanent bridge railing. A subsequent deck and joint repair project was assumed at a 15 year interval with replacement of the bridge assumed at the end of its project 75 year life. Another deck and joint repair project was assumed to occur when the new bridge was 15 years old.

2. Ceasing work on the current substantial maintenance project. The work would be renegotiated to removal of the existing rail and parapet and placing temporary concrete safety barrier in place at temporary bridge rail. The bridge would be replaced in 5 years, with subsequent deck repairs (no joints were presumed in the new design) at 15 year intervals.

The analysis is shown in Figure 6.11. As can be seen, the net present value for option one was calculated as approximately $2 million in 2015 year dollars; while the net present value for option two was calculated at approximately $2.8 million in 2015 year dollars. Option one represented significantly less cost for the study period.

With the spreadsheet, it was possible to examine the effect of changing assumptions in the timing of maintenance and replacement actions. Doing so, it was calculated that replacing the rail with permanent railing would be the most economically efficient solution for any scenario in which the bridge would not be replaced for 13 years.

As an example of the second common use of BLCCA, Afshin Hatami and George Morcous completed a study of scenarios for deck overlays, expansion joint replacements, and deck widening versus deck replacement projects for the NDOR to determine the most cost efficient strategies [19]. The analysis was conducted using RealCost, an analysis period of 60 years and construction costs from NDOR. Deterioration rates of bridge elements were taken from the deterioration rate report for Nebraska bridges by the same authors 2 years earlier [15].

LIFE CYCLE COST ANALYSIS

BRIDGE ID	0064-B0023	Route: K-177
DATE	3/18/2015	Dist. - Area - Sub: 2 - 3 - 1
ENGINEER	Paul Kulseth	County: Morris (64)
PROJECT NO.	177-64 KA-3153-01	Serial Number: 023
		Structure Type: RDGH
Reference Year	2015	State Ref. Point: 72.7
Inflation Rate	1.0%	Feature Carried: K177 HWY
Interest Rate on Investment	3.0%	Feature Crossed: MUNKER'S CREEK
Discount Rate	2.0%	Location: 6.91 MI S WABAUNSEE COLN
	http://inflationdata.com/	38.73014, -96.49938

Option 1: Maintain Existing now with Replacement in the future

ACTION	YEAR	QUANTITY	UNITS	UNIT COST (2015)	ITEM COST (2015)	DISCOUNTED COSTS	PRESENT VALUE
PATCHING (PARTIAL DEPTH)	2015	30	Sq. Yds.	$ 200	$ 6,000	$ 6,000	$ 6,000
PATCHING (FULL DEPTH)	2015	10	Sq. Yds.	$ 250	$ 2,500	$ 2,500	$ 2,500
OVERLAY (POLYMER OVERLAY)	2015	1265	Sq. Yds.	$ 57	$ 72,105	$ 72,105	$ 72,105
JOINT (NEW STRIP SEAL ASSEMBLY)	2015	62	Ln. Ft.	$ 744	$ 46,120	$ 46,120	$ 46,120
CONCRETE SURFACE REPAIR	2015	60	Sq. Ft.	$ 200	$ 12,000	$ 12,000	$ 12,000
REPLACE BEARINGS	2015	8	Each	$ 4,000	$ 32,000	$ 32,000	$ 32,000
TRAFFIC CONTROL	2015	1	LS	$ 78,615	$ 78,615	$ 78,615	$ 78,615
MOBILIZATION	2015	1	LS	$ 18,500	$ 18,500	$ 18,500	$ 18,500
CURB REPAIR	2015	100	Ln. Ft.	$ 80	$ 8,000	$ 8,000	$ 8,000
REPAIR ABUTMENT	2015	58	Sq. Ft.	$ 150	$ 8,700	$ 8,700	$ 8,700
REPLACE RAIL	2015	813	Ln. Ft.	$ 450	$ 365,850	$ 365,850	$ 365,850
OVERLAY (POLYMER OVERLAY)	2030	1265	Sq. Yds.	$ 60	$ 75,900	$ 88,118	$ 56,395
JOINT (REPAIR STRIP SEAL ASSEMBLY)	2030	62	Ln. Ft.	$ 744	$ 46,128	$ 53,553	$ 34,274
REPLACE STRUCTURE	2040	17080	Sq. Ft.	$ 125	$ 2,135,000	$ 2,737,992	$ 1,301,348
ROAD ITEMS (FOR REDECK/REPLACE)	2040	2	Each	$ 300,000	$ 600,000	$ 769,459	$ 365,719
OVERLAY (POLYMER OVERLAY)	2055	1898	Sq. Yds.	$ 80	$ 151,822	$ 226,043	$ 68,759
OVERLAY (MILL & CONCRETE OL)	2070	0	Sq. Yds.	$ 175	$ -	$ -	$ -
OVERLAY (POLYMER OVERLAY)	2085	0	Sq. Yds.	$ 80	$ -	$ -	$ -
RESIDUAL VALUE	2070	1	Each	$ (1,281,000)	$ (1,281,000)	$ (2,214,240)	$ (431,062)
					TOTALS	$ 2,311,315	$ 2,045,822

Assumptions: 1) Blue text = Costs/Quantities from Bid Tabs or Change Orders
2) Future Polymer Overlay would be a repair of the existing and only one coat with patching.
3) Future overlay unit costs are increased to include mobilization, traffic control, & patching.

Option 2: Install Barrier now, Build new structure ASAP with routine maintenance actions in the future.

ACTION	YEAR	QUANTITY	UNITS	UNIT COST (2015)	ITEM COST (2015)	DISCOUNTED COSTS	PRESENT VALUE
TEMPORARY BARRIER RAIL	2015	813	Ln. Ft.	$ 270	$ 219,510	$ 219,510	$ 219,510
REPLACE STRUCTURE	2020	17080	Sq. Ft.	$ 125	$ 2,135,000	$ 2,243,906	$ 1,933,735
ROAD ITEMS (FOR REDECK/REPLACE)	2020	2	Each	$ 300,000	$ 600,000	$ 630,606	$ 543,438
OVERLAY (POLYMER OVERLAY)	2035	1898	Sq. Yds.	$ 80	$ 151,822	$ 185,252	$ 102,172
OVERLAY (MILL & CONCRETE OL)	2050	1898	Sq. Yds.	$ 175	$ 332,111	$ 470,470	$ 166,065
OVERLAY (POLYMER OVERLAY)	2065	1898	Sq. Yds.	$ 80	$ 151,822	$ 249,692	$ 56,406
RESIDUAL VALUE	2070	1	Each	$ (711,667)	$ (711,667)	$ (1,230,133)	$ (239,479)
					TOTALS	$ 2,769,302	$ 2,781,848

Assumptions: 1) Replacement structure is 5% longer than existing with an increased roadway width
2) Replacement superstructure is weathering steel or prestressed (no paint required)
3) Replacement abutments are integral or semi-integral (no joints required)

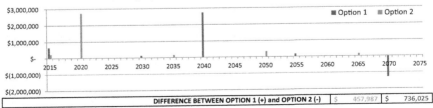

DIFFERENCE BETWEEN OPTION 1 (+) and OPTION 2 (-)	$ 457,987	$ 736,025

C:\Users\hurt\Desktop\Material for Book\Copy of LIFE CYCLE COST.xlsx

Figure 6.11 *BLCCA for Munker's Creek Bridge.*

For deck overlay, strategies involving three different types of overlay were investigated: silica fume overlays, epoxy polymer overlays, and polyester overlays. For silica fume overlays, three scenarios were compared: allow a deck to deteriorate until it reached NBI condition state 4 and then redeck, applying a silica fume overlay at condition state 6, and applying a silica fume overlay at condition state 5. It was determined that the least net present value was achieved with application at NBI condition state 6.

Neither the epoxy polymer overlay nor the polyester overlay are meant to be placed on a severely deteriorate deck. The authors assumed that each would be placed when a deck reached 15 years in service. Given the shorter service life of these two overlay types, additional applications of each would have to be made during the 60 analysis period. To match the net present value provided by applying a silica fume overlay, it was determined that an epoxy polymer overlay would require a service life of at least 14 years and a polyester overlay would require a service life of 22 years. For the overlay materials to match the net present value of leaving a bare deck until NBI condition state 4, the epoxy polymer would require a service life of 11 years and the polyester of 17 years.

Regarding policies for managing decks, the information that the BLCCA is able to provide is that applying an overlay system as a maintenance action during the life of the deck is more economically efficient than reconstructing severely deteriorate decks, that applying a silica fume overlay is better done at a higher condition rating (6 rather than 5), and that applying either epoxy polymer or polyester overlays sooner in the life of the deck and more often is an efficient strategy if sufficient service life can be derived from the materials.

REFERENCES

[1] Office of Asset Management. Life-cycle cost analysis primer. Washington, DC: Federal Highway Administration; 2002.
[2] Consumer Price Index. Bureau of Labor Statistics. [Online] United States Department of Labor. Available from: http://www.bls.gov/cpi/
[3] National Highway Construction Cost Index. Office of Highway Policy Information. [Online] Federal Highway Administration. Available from: http://www.fhwa.dot.gov/policyinformation/nhcci/desc.cfm
[4] Bridge Life-cycle Cost Analysis Guidance Manual. Washington, DC : Transportation Research Bureau. NCHRP Report 423; 2003.
[5] Mallela J, Sadasivam S. Work zone road user costs – concepts and applications. Washington, DC: Federal Highway Administration; 2011. FHWA-HOP-12-005.
[6] Hurt M.. Substantial Maintenance Practice for Bridges on the Kansas State Highway System. Lawrence : University of Kansas; 2001.

[7] Highway Capacity Manual. Washington, DC : Transportation Research Board; 2010.

[8] Jiang Y. Estimation of Traffic Delays and Vehicle Queues at Freeway Work Zones. Washington, DC: Transportation Research Board; 2001. Paper No. 01-2688.

[9] Hawk H. Bridge life-cycle cost analysis. Washington, DC: Transportation Research Board; 2003. NCHRP Report 483.

[10] Bai Y, Li Y. Determining major causes of highway work zone accidents in Kansas. Lawrence, KS: University of Kansas Center for Research, Inc; 2006. KU-05-1.

[11] Agrawal AK, Kawaguchi A. Bridge element deterioration rates. New York, NY: Transportation Infrastructure Research Consortium, NYSDOT; 2009.

[12] Agrawal AK, Kawaguchi A, Chen Z. Deterioration Rates of Typical Bridge Elements in New York. s.l. : American Society of Civil Engineering, J Bridge Eng 2010; 15 pp 419–429.

[13] Sobanjo JO. Development of Agency Maintenance, Repair & Rehabilitation (MR&R) Cost Data for Florida's Bridge Management System. Tallahassee, FL: Florida Department of Transportation; 2001. FDOT Contract BB-879.

[14] Williamson G, Weyers RE, Brown MC. Bridge deck service life prediction and cost. Richmond, VA: Virginia Transportation Research Council; 2007. VTRC 08-CR4.

[15] Morcous G, Hatami A. Developing deterioration models for Nebraska bridges. Lincoln, NE: Nebraska Department of Roads; 2011. SPR-P1(11)M302.

[16] Huang Y-H, Adams T. Analysis of Life Cycle Maintenance Strategies for Concrete Bridge Decks. New York, NY : American Society of Civil Engineers; 2004, J Bridge Eng; 9(3): pp 250–258.

[17] Krauss PD, Lawler JS, Steiner K. Guidelines for Selection of Bridge Deck Overlays, Sealers and Treatments. Wiss, Janney, Elstner Associates. Washington, DC: Transportation Research Board; 2009. NCHRP Project 20-07 Task 234.

[18] Federal Highway Adminstration. Bridge Preservation Guide. Washington, DC : s.n.; 2011. FHWA-HIF-11042.

[19] Hatami A, Morcous G. Life-cycle assessment of Nebraska bridges. Lincoln, NE: Nebraska Department of Roads, University of Nebraska; 2013. SPR-P1(12)M312.

CHAPTER 7

Bridge Management

Be it enacted by the Council and General Assembly of this state and it is hereby enacted by the authority of the same, that from and after the passing of this act, it shall not be lawful for any person or persons to drive any kind of carriage, wagon, cart, sled or sleigh, or ride any horse or mule over any of the toll bridges in this state, at a faster gait than a walk, and every person wilfully so offending, shall forfeit and pay to the proprietor or proprietors of such bridge or bridges the sum of one dollar, to be recovered by action of debt, with costs of suit in any court of competent jurisdiction.

An Act Relative to Toll and Chain Bridges, enacted on February 8, 1816 by the State of New Jersey [1]

Overview

The history and development of bridge management is discussed in this chapter. The level of service concept is defined for bridge performance and Bridge Management System (BMS) software and the principles of asset management are reviewed. The selection of projects and program development is discussed and successful practices in bridge management from Kansas Department of Transportation (KDOT) are noted.

7.1 CONTEMPORARY HISTORY OF BRIDGE MANAGEMENT

7.1.1 Before the NBIS

Laws such as the one mentioned earlier, enacted in New Jersey in 1816, used to be common throughout the United States, as were warning signs on bridges advising users of the fines for exceeding a walking pace. Until the advent of the automobile, bridges carrying public roads carried the weight of pedestrians and horse-drawn carts (Figure 7.1). Although iron came into more common use in the nineteenth century, the vast majority of bridges were made from either wood or masonry. In the United States in the nineteenth century, it was not uncommon for the maintenance on public highway bridges to be done by local landowners who were working-off poll tax obligations [2]. For the majority of bridges, bridge preservation was an activity by the local government and citizens.

With the coming of the automobile, the loading demand on bridges increased (Figure 7.2). The weights of the first cars were comparable to those of

Highway Bridge Maintenance Planning and Scheduling
http://dx.doi.org/10.1016/B978-0-12-802069-2.00007-6

loaded wagons, but by 1913 there were already four states in the United States enacting weight limits for trucks on public roads. The heaviest of which was 28,000 pounds in Massachusetts [3].

With the mobility afforded by the automobile, the expectation of the public for accessibility on all of the public roads increased. In the present day, a parcel of property which cannot be driven to in almost any weather condition in an automobile is considered inaccessible. The result was the expansion of the inventory of bridges on public roads in the twentieth century. These bridges were designed to carry greater loads than their predecessors

Figure 7.1 *The Kansas River bridge at Lawrence, Kansas 1867. (Courtesy of George Allen and the Douglas County Historical Society, Watkins Museum of History).*

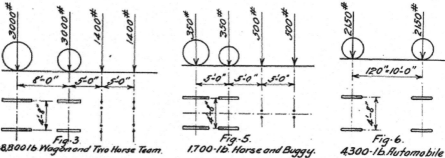

Figure 7.2 *Moving Loads Used in Testing a Truss Bridge by the University of Illinois in 1913 [4].*

and were constructed from materials such as reinforced concrete and steel. Maintenance activities on these bridges require heavier equipment and expertise that is beyond the everyday experience of the landowners who had maintained public bridge previously.

With the growing inventory of highway bridges the need for bridge maintenance also grew. As discussed in Chapter 1, high-profile bridge failures in the late 1960s led to the development of the NBIS. Title 23, Part 650 of the Code of Federal Regulations requires states to maintain an inventory of all bridge structures. Through the 1970s bridge inspection, data was accumulated which allowed bridge owners to see the condition of their bridges through time and to make comparisons across the nation's inventory of bridges. Eligibility for federal funds at this time through the Federal Highway Administration (FHWA) for bridge work under the Highway Bridge Replacement and Rehabilitation Program and the Special Bridge Program was based on National Bridge Inventory (NBI) condition and appraisal ratings. Decisions on which bridges would have maintenance or rehabilitation work were made on the basis of inspection data.

7.1.2 Definition of Bridge Management

The simplest definition for bridge management is that it is the activity of administrating resources to maintain operational bridges. Over several decades, this scale of this activity has increased from managing a few bridges with local labor to making decisions on the state and federal level considering and comparing thousands of bridges at a time. By the 1980s there was significant interest at the state and federal levels in developing efficient and effective bridge management systems (BMS).

In contemporary discussions, BMS is often conflated with software. Although software is required to handle the large amounts of data involved in making decisions for inventories of bridges numbering in the thousand, the software is only a tool. Paraphrasing the FHWA report, *Bridge Management Systems*, a BMS is a system or series of engineering and management functions which, taken together, comprise the actions necessary to administrate resources to maintain an inventory of bridges in operation [5]. In practice, typically a number of software applications are utilized to provide the manager with information to further process and make decisions based on engineering judgment.

The flowchart from the 1989 report, remains one of the most complete representations of the complete bridge management process in the literature (Figure 7.3). The first step is bridge inspection information. This data is at the heart of any BMS. It is combined with information on construction projects

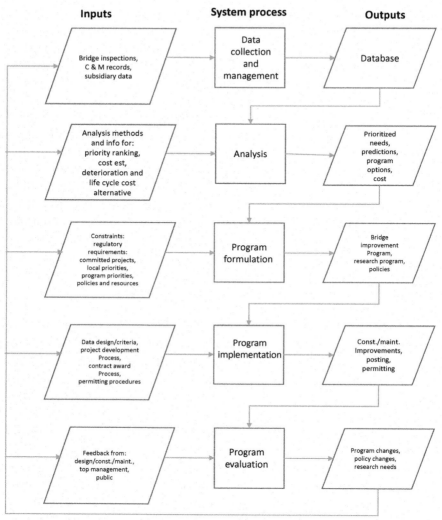

Figure 7.3 *Flowchart of Bridge Management Activities. (Adapted from Ref. [5]).*

and maintenance activities and associated with specific bridges in a database. In the second activity, information used to conduct an analysis of bridge needs to form a list of prioritized bridge needs and estimated costs to address them. The third activity is to choose projects from among the needs and form a maintenance program. Programs are usually complied on a fiscal year basis, although in times of high needs and low resources programs may be constantly updated to match need with resources as they are available. The fourth activity is to implement the maintenance program and any other actions, such as load posting that are required. The fifth activity is to then evaluate the result of maintenance actions with feedback from all involved parties. Any changes to policies or practice then are reflected in the next iteration of the process.

The activities of the bridge management process are used to direct the bridge preservation cycle of inspection, evaluation, and maintenance. The problems of particular bridges are evaluated in the second activity. But prior to selection maintenance projects for inclusion in a program in activity three an evaluation of how the bridge work will fit into system-wide priorities should occur. As bridges are a part of the larger transportation network, a BMS is a part of a larger asset management process. In evaluating priorities for a highway system there are bridge characteristics that are not related to the structural condition of bridge elements, such as roadway width, which are significant due to their effect on traffic operations.

7.1.3 Level of Service

Among the first work done to support modern BMS methodologies was the development of level of service goals in a research project at North Carolina State University for the North Carolina Department of Transportation [6]. They selected three characteristics as measures for bridge improvement needs: load capacity, clear deck (roadway) width, and vertical roadway clearance. Each characteristic was assigned an acceptable level of service goal and a desirable level of service goal. Acceptable levels of service were meant to meet the needs of the majority of expected vehicles on the route. Desirable levels of service met the standard of new construction at that time. The level of service goals from the report are shown in Tables 7.1–7.4.

With these measures for level of service provided for each bridge, it can be determined whether a bridge is fulfilling its function on the highway system. A bridge in poor structural condition which is also providing a low level of service might better be replaced than repaired. Establishment of level of service criteria also allowed for priority ranking of bridges as discussed in Section 7.2.1.

Table 7.1 NCDOT bridge load capacity goals from Johnson and Zia [6]

Functional classification of road carried	Single vehicle capacity (tons)	
	Acceptable	Desirable
Interstate and arterial	Not posted (= 33.6 tons)	Not posted
Major collector	25	Not posted
Minor collector	16	Not posted
Local	16	Not posted

Table 7.2 NCDOT bridge vertical clearance goals from Johnson and Zia [6]

Functional classification of road under bridge	Single vehicle capacity (ft.)	
	Acceptable	Desirable
Interstate and arterial	14.0	16.5
Major and minor collector	14.0	15.0
Local	14.0	15.0

Table 7.3 NCDOT clear deck width goals for two lane routes from Johnson and Zia [6]

Functional classification of road carried	Current ADT	Clear width (ft.)	
		Acceptable	Desirable
Interstate and arterial	800 or under	22	24
	801–2000	24	36
	2001–4000	26	40
	Over 4000	28	40
Major and minor collector	800 or under	20	24
	801–2000	22	28
	2001–4000	24	30
	Over 4000	26	30
Local	800 or under	20	24
	801–2000	22	28
	2001–4000	24	30
	Over 4000	26	30

7.1.4 BMS Software

At this time several other states were researching and developing aspects of BMS [7]. In 1985, National Cooperative Highway Research Program (NCHRP) Project 12-28 [2], Bridge Management Systems Software, was initiated to develop a model form of BMS software for the network level of analysis. The concepts developed in the first phase of this project were documented in NCHRP Report 300, *Bridge Management Systems*. This research continued and resulted in the release of the program Bridgit in 1998 with NCHRP Project 12-28(c).

Within Bridgit there were five basic modules [8]:

1. Inventory – a database allowing for user-defined data items (such as vertical clearance) to be associated with each bridge or element of a particular bridge;

2. Inspection – a module to view or edit element-level bridge inspection information and to view network-level historic information;

Table 7.4 NCDOT clear deck width goals from Johnson and Zia [6]

Functional classification of road carried	Current ADT	Acceptable (ft.)		Desirable (ft.)	
		Lane	Shoulder	Lane	Shoulder
Interstate and arterial	800 or under	10	1	12	4
	801–2000	10	2	12	6
	2001–4000	11	2	12	8
	Over 4000	11	3	12	8
Major and minor collector	800 or under	9	1	10	2
	801–2000	9	2	11	3
	2001–4000	10	2	12	3
	Over 4000	10	3	12	3
Local	800 or under	9	1	10	2
	801–2000	9	2	11	3
	2001–4000	10	2	12	3
	Over 4000	10	3	12	3

3. Maintenance, rehabilitation, and replacement – a module to track maintenance activities on individual bridges and across the network of bridges, with routines to provide cost estimates for work;
4. Analysis module – this is capable of producing network-level work plans under defined annual budgets to meet goals for improvement of condition or of defined level of service; and
5. Models module – this module permits user modification of models and tables in the other four modules.

Bridgit was one of the first comprehensive BMS software implementations. It performed the functions envisioned for a comprehensive BMS software application:

- Keeping information about an inventory (or network) of bridges in a database where it could be readily retrieved and where it could be used to generate statistics indicative of the condition of the inventory;
- The ability to project the future condition of bridges based on their current condition and the application of deterioration modeling;
- A capacity to project maintenance costs given projections of future condition and rules about timing and cost of maintenance actions; and
- The ability to process multiple scenarios of maintenance activity to allow for optimization.

Maintenance of the software was taken over by the American Association of State Highway and Transportation Officials (AASHTO) in 1998. It appears to have fallen out of use with development of other AASHTOWare products.

In 1989, the FHWA issued a report on BMS as part of Demonstration Project 71, which examined and documented state of the art practices and research at the time [5]. The FHWA project combined with the effort of several states, including California and Florida, continued with the development of the Pontis bridge management system. Pontis and Bridgit were similar systems with the primary difference being the optimization routine. Optimization of proposed work plans is made with bottom–up approach in Bridgit as opposed to a top–down approach in Pontis. In the bottom–up approach individual bridge projects are determined and then aggregated to form a system–wide program, which is compared to the available budget. The top–down approach starts with the budget and goals and then refines down in individual bridge projects. The disadvantage of the bottom–up approach is that the system for Bridgit is slower than Pontis for larger bridge populations; an issue more significant on the personal computers of the late 1990s than it would be now.

In 1991, the US Congress passed the Intermodal Surface Transportation Efficiency Act (ISTEA) [Public Law 102-240], which mandated the development and implementation of intermodal management systems by states. This included BMS. Even after the requirements for adoption of a BMS were repealed with the National Highway System Designation Act [Public Law 104-59] in 1995, development of BMS software still continued.

Maintenance for Pontis was also taken over by AASHTO. By 2009, 44 states were licensing Pontis for use from AASHTOWare[1] [9]. The remaining states were using BMS software particular to their state. Although Pontis had continued to improve its deterioration modeling and added functionalities to the program, a NCHRP study of the use of BMS software by transportation agencies found that completely half of the Pontis users used only the inspection database capabilities [10]. Only 40% of the users made any use of the programming module for network analysis of bridge needs. Often agencies were too constrained by limited maintenance budgets to pursue plans of work that would lead to reduced life cycle costs but at the expense of increased project funding now. Other agencies, though, considered the wider implications of bridge projects, including operational impacts and the criticality of needs.

7.1.5 Asset Management

At the same time as BMS software was becoming capable of a greater degree of mathematical analysis, a trend toward incorporating asset

---

[1] In the years since Pontis has become AASHTOWare Bridge Management, part of a suite of programs including AASHTOWare Bridge Design and AASHTOWare Bridge Rating [10].

management principles to the management of infrastructure was growing. In asset management, performance goals and measures are identified and policies are adjusted to meet the goals. From NCHRP Report 551, *Performance Measures and Targets for Transportation Asset Management*, the core principles of asset management are [11]:

- Policy driven – Resource allocation decisions are based on a well defined and explicitly stated set of policy goals and objectives. These objectives reflect desired system condition, level of service, and safety provided to customers and are typically tied to economic, community, and environmental goals.

- Performance-based – Policy objectives are translated into system performance measures that are used for both day-to-day and strategic management.

- Analysis of options and tradeoffs – Decisions on how to allocate resources within and across different assets, programs, and types of investments are based on understanding how different allocations will affect the achievement of policy objectives and what the best options to consider are. The limitations posed by realistic funding constraints also must be reflected in the range of options and tradeoffs considered.

- Decisions based on quality information – The merits of different options with respect to an agency's policy goals are evaluated using credible and current data. Decision support tools are applied to help in accessing, analyzing, and tracking these data.

- Monitoring to provide clear accountability and feedback – Performance results are monitored and reported for both impacts and effectiveness. Feedback on actual performance may influence agency goals and objectives, as well as future resource allocation and use decisions.

With a focus on setting policy to meet performance goals, asset management approaches differ from both the traditional approach of addressing the worst bridges first and straight optimization of condition for least cost across the entire inventory.

In a case study of bridge management practices in Idaho, Michigan, and Virginia, the FHWA found that the states employ bridge preservation as a means of bridge management [12]. The report states:

Structures in fair or good condition that have adequate traffic capacity and adequate load rating are preserved more easily than they could be rebuilt or replaced. But needs at structures in poor condition may seem to be more urgent. Funding, always limited, can be exhausted in projects for structures in poor condition if priorities are evaluated among structures in a single population. These states compensate by segregating bridges by condition; either good, fair or poor. Bridges in good condition are candidates for maintenance actions, those in fair conditions may receive repair, and the poor condition bridges are eligible for rehabilitation work. The key is that there are separate funds for each level of work, so that cost effective investments to maintain bridges in good condition are still made.

With the passage of The Moving Ahead for Progress in the 21st Century (MAP-21) in July 2012, the Secretary of Transportation was required to promulgate a rule to establish performance measures in specified federal aid highway program areas including bridges. As of 2015, the FHWA was still involved in the process of the Notice of Proposed Rulemaking, where the public's comment is sought. At that time, the two proposed performance measures to track a state's bridge inventory are the percentage of National Highway System (NHS) bridges in Good Condition and the percentage of NHS bridges in Poor Condition. The assessment of whether a bridge is in good or poor condition is the following lowest of the following condition ratings for the NBI Items: deck, superstructure, substructure, or culvert. The rating of each bridge would be weighted by the deck area of the bridge and the condition totals for the state would be expressed as a percentage of the total bridge deck area for the state [13].

7.2 PROJECT AND PROGRAM SELECTION

7.2.1 Priority Ranking

Despite the ever-growing capabilities of BMS software to analyze the condition data of entire bridge inventories and produce recommendation for maintenance project, the vast majority of bridge projects are selected by producing a prioritized list of proposed projects from candidate structures screened for need. The ability to query and screen bridges in the BMS database for particular characteristics or conditions makes a simple process. While the optimization routines of BMS software may appear as a black box to all but the most experienced user, the methods used in priority ranking are easily understood and transparent.

The FHWA Bridge Management Systems report describes three prototypical processes used to establish priority rankings for projects [5]. The simplest process involves an initial screening of bridges with set criteria to establish need, then conducting a review of the list and selecting and scoping projects on the basis of engineering judgment. The next more complex process involves an initial screening of bridges, and then determining if

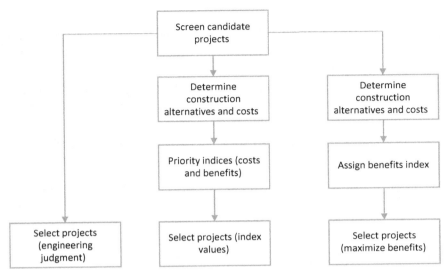

Figure 7.4 *Prototypical Processes for Setting Program Priorities. (Adapted from Ref. [5]).*

there are construction alternative to consider investigating for programming. A priority index (PI) is calculated for likely alternatives. Projects with the highest PIs are chosen, subject first to passing review by engineers (Figure 7.4). The third prototypical process is similar, except a benefit index is calculated which may include savings in life cycle costs or net benefits to users.

The third process is rarely used for bridge work, except in the case of complex bridges. It is used for higher costs projects where a benefit/cost analysis may be required to make the case for investment. The second process is common for the selection of bridge replacement and large rehabilitation projects. It involves the use of a PI formula, which may be used to allow consideration of factors other than a bridge's structural condition in project selection. The first process is how most maintenance projects in the United States are programmed.

7.2.2 Programming Maintenance Work

The goal of maintenance work is to restore a bridge or particular bridge elements to an acceptable condition. Rehabilitation work, such as redecking, may be engaged to improve the functionality of a bridge; but screening for maintenance and repair projects should be limited to condition criteria. For maintenance, bridges should be screened with the intent of finding candidates that would benefit from maintenance actions.

There are a number of measures in the NBI inspection that reflect whether a bridge is in a damaged or deteriorated condition. The most obvious is whether a bridge is classified as deficient, that is either structurally deficient (SD) or functionally obsolete (FO), or not. Note that a bridge may be classified as FO if it receives an appraisal rating of three or less on a number of roadway items, or on structure evaluation; that is FO may or may not be reflective of impaired condition.

A bridge will be classified as SD if any of the following NBI condition ratings is a 4 or less, deck, superstructure, substructure, or culvert. An initial screen of bridges for candidates would probably choose to include those with a somewhat higher NBI rating to address condition before the structure is deficient.

With the adoption of element-level inspections, information on the condition of problematic elements, such as expansion joints, is available from the bridge inspection reports. A reviewer may choose to query for bridges with any portion of an element in severe condition, or for elements with a certain percentage in poor or severe conditions. Whereas a deteriorating expansion joint or a single damaged girder end might not affect the overall condition of a component enough to significantly drop an NBI rating, the finer detail of element level inspection captures that information in the inspection database.

No matter how well crafted a database query or how well screening criteria may appear to be crafted, bridges that ought to be addressed with a maintenance action appear to slip through the process. For bridges with deficiencies that require replacement, rehabilitation, or redecking, bridge inspectors are required to make note in NBI item 75 of the proposed scope. Inspectors are also required to report Critical Findings (see Section 3.2.5), which should be addressed as soon as possible to preserve either structural integrity or traffic safety. Good inspection practice; however, should require inspectors to make note of any proposed maintenance recommendations in the bridge inspection report. Maintenance recommendation should be sorted by whether they would entail substantial maintenance (contracted) or preventative maintenance (owner's forces) work. If this is in place and database has a field to make note of whether maintenance has been recommended for each bridge, then the initial screening for candidate work is all but complete.

An initial screening of an inventory of bridges to determine maintenance candidates by need, then entails screening the current bridge inspection report of each bridge for:
- NBI condition ratings of a selected level or less;
- Element ratings reporting a selected percentage of poor or worse condition; and
- Inspector-recommended maintenance.

A bridge does not have to be in dire condition to benefit from maintenance. It is often bridges which are in fair condition for which actions that maintain condition will have the most benefit and result in lower life cycle costs. To this end, it is imperative to have established maintenance policies that have been determined to result in lower life cycle cost before determining criteria to use to select projects. An example of such policies was discussed in Section 6.5.2 where the Nebraska Department of Roads (NDOR had determined through a Bridge Life Cycle Cost Analysis (BLCCA) that the most cost-effective strategy for applying deck overlays was to either apply a silica fume overlay when the deck condition reached a NBI condition rating of 6, or to apply either an epoxy-polymer overlay or polyester overlay when the bridge reached the age of 15 years and to reapply afterward at specific intervals. A Nebraska engineer screening bridges may choose to list all span bridges with a deck condition rating of 6 or less and/or bridges 15 years of age or older without an initial overlay.

It is the application of protective systems, such as deck overlays or paint, for which a strategy of applying before the beginning of gross deterioration starts may have the best life cycle cost impact. In addressing deterioration that has already begun, such as corrosion at a girder end; if the deterioration is addressed before compromise of the condition of the component, it may be of benefit in terms of reduced future costs to address it.

An initial screening of an inventory of bridges to determine maintenance candidates by potential benefit to life cycle cost then entails screening the current bridge inspection report of each bridge containing elements for which preventative maintenance strategies have been identified, for specific NBI component condition ratings and element ratings. The trigger value for the ratings will be higher than those indicative of current need.

Once a list of candidates for maintenance work has been generated it is typically reviewed by experienced engineer with knowledge of the bridges and standard practices for the owner. Candidates due to need may initially be ranked by the severity of deficiencies while candidates due to benefit may be ranked by the scale of potential benefit (e.g., overlaying a bridge with a large deck area early may accrue more life cycle cost savings than overlaying a smaller bridge early). Candidate bridges are then vetted for work utilizing the four criteria for initial selection discussed in Section 6.1.1:

1. Is there a deficit that may be economically addressed by maintenance work?
2. Is this bridge necessary?
3. What is the minimum condition necessary for the operation of this bridge?
4. What is the likely remaining service life of this bridge?

With the list refined to those bridges which would benefit from maintenance work, initial project scopes are developed based on the maintenance policies and strategies adopted by the bridge owner. The project costs are estimated and compared to the available budget. If the proposed expenditures exceed the budget, the scopes are re-reviewed for work utilizing the criteria for determining scope discussed in Section 6.1.2:

1. What is the desired condition for this bridge?
2. Is there a need for similar bridge work in the area?
3. Where does this bridge weigh against others on the basis of need?

As noted in the best practices from asset management, some funding should be set aside and dedicated to projects based on benefit to long-term life cycle costs. This should be seen as an investment to be made whenever possible, though tight budgets may force owners to address only the worse of identified needs.

7.2.3 Priority Indexing

To consider a bridge for replacement or rehabilitation it must either be in such a deteriorated condition that repair is not an economical option, or have functional shortcomings (e.g., narrow roadway) that must be addressed for operational concerns for the route. The screening and the prioritizing of bridges for replacement or rehabilitation typically involve a composite measure of several attributes. This measure is usually taken in the form of a PI.

A PI has the general form of Ref. [5]:

$$PI = \sum_i k_i f_i (a,b,c\dots)$$

Where, PI, priority index; k_i, weighting factor of the ith objective function $(\sum k_i = 1.0)$; $f_i = i$th, objective function; and $a, b, c\dots$, values of attributes included in the objective function.[2]

The objectives and the weight given to each objective should reflect the policies and goals of the bridge owner. As such, PIs may also be performance measures for the owner. If an owner is concerned solely about structural condition, a PI which is a sum of measures of condition only would be apropos. If the owner is concerned about narrow bridges or insufficient vertical clearance, those items could be included and appropriately weighted in the PI.

---

[2] In discussion of the PI formula, the reference cited refers to "need" functions rather than "objective" functions. The change has been made in this text for clarity of the discussion.

A PI readily available from the bridge inspection files is the sufficiency rating (discussed in Section 3.2.4) for each bridge. The sufficiency rating is the sum (ranging from 0 to 100) of the performance of the bridge in regards to four objectives: S_1, structural adequacy and safety; S_2, serviceability and functional obsolescence; S_3, concerns for public use; and S_4, special reductions based on detour length, traffic safety features, and structure type. Maximum values of the objective functions are: $S_1 = 55$, $S_2 = 30$, $S_3 = 15$, $S_4 = -13$ (S_4 is a deduction only and the minimum value of the sufficiency rating is limited to zero).

Sufficiency ratings have traditionally been used as a determinate of eligibility for federal funding for the FHWA Highway Bridge programs. It places approximately equal value on the load capacity and structural condition (S_1) as the functional and traffic network characteristics (S_2, S_3, and S_4) of the bridge. While it is a useful measure, bridge owners have found it insufficient for use in prioritizing bridges for programming.

On one hand, it gives a smaller weight to bridge condition than might be desired if the goal is to minimize life cycle costs. A long span bridge with a wide roadway and good load-carrying capacity would have a high-sufficiency rating even if its deck proved to be a source of constant maintenance issues. On the other hand, a load posting would significantly drop the sufficiency rating even if it were simply a lightly designed bridge in good condition on a route where the posting is not an issue.

One approach to resolve the shortcomings listed has been to develop a composite measure of the condition of a bridge to use for screening in conjunction with a state-specific priority formula to prioritize candidates for replacement or rehabilitation.

A composite measure of bridge condition developed by the California Department of Transportation, and subsequently adopted by several states, is the bridge health index (BHI). The BHI is the ratio of the sum of the current value of elements to sum of the initial value of elements for all elements in the bridge [14]. The initial value of an element is the quantity of the element times the failure costs of the element. For the current value of the element, it is weighted by the quantity of element in each of the element level inspection condition states. See the calculation[3] of BHI for a sample bridge in Tables 7.5 and 7.6.

$$\text{For this example, BHI} = \frac{\sum \text{CEV}}{\sum \text{TEV}} \times 100 = \frac{\$730,626}{\$762,800} \times 100 \approx 0.96$$

[3] The example calculation has been adopted and modified from the referenced article [14].

Table 7.5 Element distribution for BHI calculation

Element	Units	Quantity	Condition state 1 (100%)	Condition state 2 (67%)	Condition state 3 (33%)	Condition state 4 (0%)	Unit failure cost ($)
Concrete deck	Square ft.	3040	2500	540			70
Steel girders	Ft.	320	300		20		1000
Elastomeric bearings	Each	8	8				1500
Reinforced concrete abutment	Ft.	100	100				1800
Strip seal expansion joint	Ft.	76	50	20		6	500

Assumed bridge, 36 ft.; roadway, 80 ft.; single span, steel beam.

Table 7.6 Calculation of TEV and CEV for BHI

Element	Calculation of TEV	TEV ($)	Calculation of CEV	CEV ($)
Concrete deck	3040 × 70	212,800	[2500 + (540 × 0.67)] × 70	200,326
Steel girders	320 × 1000	320,000	[300 + (20 × 0.33)] × 1000	306,600
Elastomeric bearings	8 × 1500	12,000	8 × 1500	12,000
Reinforced concrete abutment	100 × 1800	180,000	100 × 1800	180,000
Strip seal expansion joint	76 × 500	38,000	[50 + (20 × 0.67)] × 500	31,700
	ΣTEV =	762,8000	ΣCEV =	730,626

TEV, total element value (initial state); CEV, current element value.

The BHI is useful as a screening tool and as a performance measure of condition. KDOT has adopted the following definitions for overall bridge condition relative to BHI:
- Good Condition: BHI > 85,
- Fair Condition: 70 < BHI ≤ 85, and
- Deteriorated Condition: BHI ≤ 70.

As a performance goal, KDOT aims to have 85% or more of the state bridge inventory in good condition and less than 5% in deteriorated condition [15]. As a screening tool for potential bridge replacement or rehabilitation candidates, one might choose to advance only deteriorated bridges or to advance bridges that are either fair of deteriorated as candidates for work.

There is no one-priority formula for ranking candidates for replacement or rehabilitation work. Each priority formula represents the values, conditions, politics, and transportation policies unique to each state or owner. All bridge owners, however, have some common interest. They desire to maintain their inventories in good structural condition to minimize life cycle costs. They desire to promote user safety and they desire to maintain good functionality on the system. The particular measures that they choose and the weights that they assign to these goals are individual to each owner. The development of the priority formula used by Kansas to select bridge replacement and rehabilitation projects is discussed as an example.

In 1980, at the direction of the State Legislature, KDOT began development of a priority ranking procedure for construction of all road and bridge projects on the State Highway System [5]. To develop this procedure the Delphi process was employed. The Delphi process was developed by the RAND Corporation in the 1950s to determine the consensus opinion of experts. The process entails assembling a group of experts (for KDOT there were 25 individual participating) who anonymously reply to inquiries concerning the subject at hand. Subsequently, the groups receive a summary of the responses and the inquiries are repeated. The intent is for the experts to converge and develop a consensus reflective of the experience of the group [16].

In the process of developing the ranking procedure, objectives were identified. For bridges the objectives were to maximize user safety, maximize preservation of investment and to minimize user travel time, and operational costs.

To assess user safety, it was chosen to examine driver exposure. At a bridge site, it was decided that driver exposure was measured by both the horizontal clearance to obstructions and the roadway restrictions caused a bridge being narrower than the approaching roadway. The driver exposure was adjusted by the historical accident rate at the site.

To assess preservation of investment the deck NBI condition rating and the lesser of the superstructure and substructure NBI condition ratings were used.

To assess user travel time and vehicle operation costs the proxy measure of operating rating was chosen. Low operating rates correlate with load postings resulting in inconvenience to commercial and agricultural traffic.

After the measures for the objectives were selected, the Delphi process was used to establish midpoint values for each and what tradeoffs were between the objectives. The midpoint was used to establish functions that provide a value between 0 and 1 for each measure, for example, a NBI deck rating of 4 resulted in a value of 0.55 use for deck condition in the priority formula (Figure 7.5). The tradeoff was used to establish the weight assigned to each measure, shown in Table 7.7.

The priority formula also included five adjustment factors:
1. Functional classification, from 1.0 for interstates to 0.60 for minor collectors;
2. Traffic volume, from 0.381 for Average annual daily traffic (AADT) = 0 to 1.0 for AADT = 10,000;
3. Bridge Adjustment Factor of 0.53 to account for the number of attributes given in the policy for bridges;
4. Accident rate, from 0.734 for low rates to 1.0 for high rates; and
5. Posted speed limit, from 0 at 5 mph to 1.0 at 55 mph.

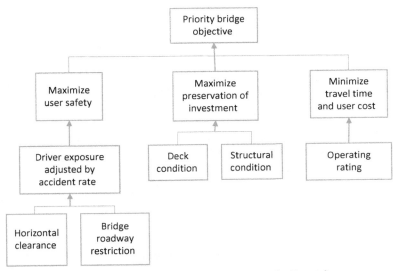

Figure 7.5 *Objectives and Attributes for KDOT Bridge Priority Formula.*

The priority formula developed was:

$$PI = AF_{Br} AF_{FC} AF_{Tr} \left(AF_{PS} AF_{Ac} \right) \left(0.196H + 0.088\,BR \right) + \left(0.232\,DC \right) \\ + \left(0.314\,SC \right) + \left(0.170\,OR \right)$$

Where, AF_{Br}, bridge adjustment factor; AF_{FC}, functional classification adjustment factor; AF_{Tr}, traffic volume adjustment factor; AF_{PS}, posted speed adjustment factor; AF_{Ac}, accident rate adjustment factor; H, value for horizontal clearance function; BR, value for bridge roadway restriction function; DC, value for deck condition function; SC, value for structural condition function; and OR, value for operating rating function.

The higher the PI value, the greater the need for work. Note, the formula has been updated in the decades since, but still retains the general form and similar weights. The KDOT formula is provided as an illustration of the

Table 7.7 Weights assigned in original KDOT priority formula [5]

Measures	Weight
Horizontal clearance	0.196
Bridge roadway restriction	0.088
Deck condition	0.232
Structural condition	0.314
Operating rating	0.170

concept. Each state and bridge owner will have their own prioritization process for repair and rehabilitation work.

7.3 EXPERIENCES FROM THE KANSAS DEPARTMENT OF TRANSPORTATION

Although much of the referencing of practice by KDOT in this text is due to the first authors' familiarity with the agency's practice, it is also due to the agency's performance in bridge management and preservation. In 2010, KDOT was chosen, along with three other state departments of transportation, for review in NCHRP Project 20-24 for best practices in bridge condition [17]. This was due to strong performance in improvement of NBI ratings from 1999 to 2009 and for the strength of 2009 ratings.

This performance is the result of KDOT abiding by two general principles concerning bridge preservation and bridge management in the years of and the years prior to 1999–2009 reviewed. They are:
1. An intentional commitment to bridge preservation activities; and
2. Close coordination of bridge inspection with bridge design and KDOT field personnel.

7.3.1 Commitment to Bridge Preservation

KDOT has made use of advances in bridge management procedures and activities as they have become available. It was early adopter of Pontis and element level inspection, in 1993, and has used BHI has a performance measurement to track performance measures since the late 1990s. More importantly, though, KDOT has commitment funding specifically to bridge preservation since 1990.

In 1989, Kansas enacted an 8-year highway program with a focus on improving what had been a deteriorated state highway system. As part of the changes with the new program, the position of Bridge Management Engineer was created in 1990 to oversee bridge inspections and evaluations. This position also was assigned to be the program manager for bridge substantial maintenance funds. These were funds set aside in the multiyear highway program for projects to be selected annually by the bridge management engineer on the basis of bridge inspection results.

The success of the first highway program led to passage of a 10-year program in 1999. Funding was increased for Bridge Substantial Maintenance from an annual average of $7.4 million during the years 1990–1997, to $14.9 million during the years 2000–2009. With the new highway program, two new funds under the authority of the Bridge Management Engineer were established for bridge rehabilitation: Bridge Redeck and Priority Bridge Culvert replacement. Together the two new funds averaged $5.3 million annually.

7.3.2 Coordination Between Bridge Inspection and Design

With centralized bridge inspection and the establishment of the bridge management engineer it was soon realized that programming maintenance work from that group was an effective way to efficiently allocate maintenance funds. Not only was the fund manager privy to the most current and complete picture of bridge needs, but the effectiveness of particular maintenance actions was communicated directly back to the fund manager in subsequent inspection cycles.

In 1999, a bridge maintenance plans squad was created to facilitate the increased focus on preservation activities with the increased funding of the second multiyear highway program. The unit acted as a bridge design squad specializing in bridge maintenance and rehabilitation work. With it able to work very closely with bridge inspection, the effectiveness of various maintenance and repair procedures could be shared directly with designers. Additionally, the performance of previous bridge designs and types of construction was shared directly with in-house bridge designers. Problematic details were eliminated from new designs, resulting in a bridge inventory that was increasingly comprised of more maintainable bridges.

The close communication and coordination facilitated by committing to bridge preservation in the structure of the organization itself, did more to improve performance in maintain bridge condition than any individual process or technique adapted before or since.

REFERENCES

[1] Thomas Edision State College. New Jersey Session Laws. New Jersey State Library. [Online] Available from: http://www.njstatelib.org/research_library/legal_resources/historical_laws/session_laws/
[2] Schirmer S, Wilson T. Milestones: a history of the Kansas Highway Commission & the Department of Transportation. Topeka: Kansas Department of Transportation; 1986.
[3] U.S. Department of Transportation. Comprehensive truck size and weight study. Washington, DC: Federal Highway Administration; 2000. FHWA-PL-00-029.
[4] Dufour FO. Some experiments on highway bridges under moving loads. J Western Soc Eng 1913;XVIII:554–79.
[5] O'Connor D., Hyman WA. Bridge management systems. Washington, DC: Federal Highway Administration; 1989. FHWA-DP-71-01R.

[6] Johnston DW, Zia P. A level of service system for bridge evaluation. Raleigh, NC: North Carolina State University, Center for Transportation Engineering; 1984.

[7] Small EP, Philbin T, Fraher M, Romack GP. Current status of bridge management system implementation in the United States. Washington, DC: Transportation Research Board; 1999. Transportation Research Circular 498.

[8] Hawk H. BRIDGIT: user-friendly approach to bridge management. Washington, DC: Transportation Research Board; 1999. Transportation Research Circular 498.

[9] AASHTO. BRIDGEWare Update Newsletter; 2009. vol. 3.

[10] Markow MJ, Hyman WA. Bridge management systems for transportation agency decision making. Washington, DC: Transportation Research Board; 2009. NCHRP Synthesis 397.

[11] Systematics, Cambridge. Performance measures and targets for transportation asset management. Washington, DC: Transportation Research Board; 2006. NCHRP Report 551.

[12] Hearn G. Bridge management practices in Idaho, Michigan and Virginia. Washington, DC: Federal Highway Administration; 2012. Transportation Asset Management Case Studies.

[13] Federal Highway Adminstration. Office of Transportation Asset Management Fact Sheet – pavement and bridge condition performance measures. Washington, DC: Federal Highway Administration; 2015. FHWA-HIF-15-003.

[14] Shepard RW, Johnson MB. California Bridge Health Index: A Diagnostic Tool To Maximize Bridge Longevity, Investment. 215, Washington, D.C. : Transportation Research Board; July–August 2001, TR News, 6-11.

[15] Fleck T. Bridge Office Presentation at the 2010 Bridge Design Workshop. Manhattan, KS: Kansas State University; 2010.

[16] Helmer O. Analysis of the future: the Delphi method. Santa Monica, CA: RAND Corporation; 1967.

[17] Spy Pond Partners, LLC. Measuring Performance Among State DOT's Comparative Analysis of Bridge Condition Final Report; 2010. NCHRP 20-24(37)E.

[18] AASHTOWare Bridge Management. AASHTOWare. [Online] Available from: http:// aashtowarebridge.com/

APPENDIX 1

Delay Calculation for Undercapacity Flow at a Typical Signalized Work Zone for Bridge Deck Repair Work

Capacity has been estimated for the signalized lanes of the work zones of a typical bridge deck repair as half of 1400 vph, the capacity of a lane through a work zone specified in the 2010 Highway Capacity Manual. With that assumption, the v/c ratio for the peak hour volumes is less than 0.6 for projects at bridge sites with an Annual Average Daily Traffic (AADT) of less than 8000 (where peak hour volume is less than 9% of AADT).

Delay will be calculated for a typical work site and the delay time will be applied to all the signalized bridges in this appendix. Delay will be assumed to consist of an average delay of the controller plus delay from the vehicle traveling the work zone at reduced speed.

The purpose of this appendix is to determine a conservative value for work zone delay for preliminary review of signalized work zones for bridge deck repair. Signals at work zones may be actuated, but reasonable signal timing will be estimated to conservatively determine delay.

Since the lanes are well under capacity at peak hours, assume uniform delay to estimate delay from the controller.

The work zone traffic control is assumed to follow Kansas Department of Transportation (KDOT) Traffic Engineering Standard TE732 for Temporary Signalized Work Zones.

Step 1 – Determine typical layout of work zone.

Step 2 – Determine typical signal timing.

Step 3 – Calculate an average uniform delay from the signal controller.

Step 4 – Calculate the delay in traveling through the work zone for four cases:

1. Passenger car with no stop.
2. Semitrailer truck with no stop.
3. Passenger car with stop.
4. Semitrailer truck with stop.

Highway Bridge Maintenance Planning and Scheduling
http://dx.doi.org/10.1016/B978-0-12-802069-2.00008-8

Step 5 – Calculate total average delay for each passenger car and semi-trailer

STEP 1

Referring to Standard TE732 for layout.

Values for dimensions *A*, *B*, and *C*, from TE710 for Rural Highways: *A* = 750 ft., *B* = 750 ft., *C* = 750 ft.

Buffer = 165 ft. for original site speeds > or = 60 mph.

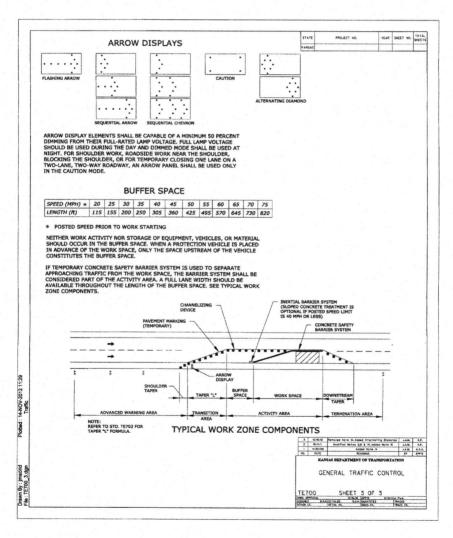

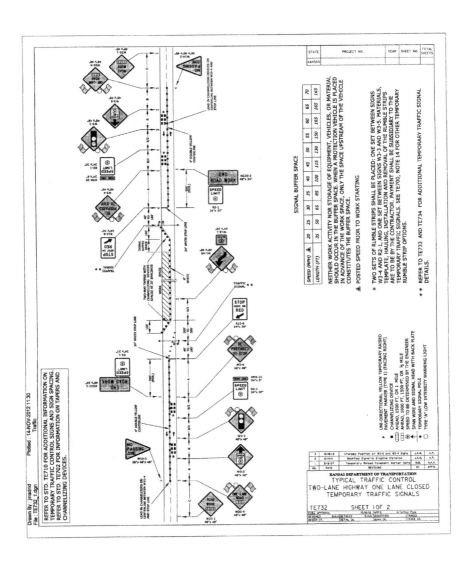

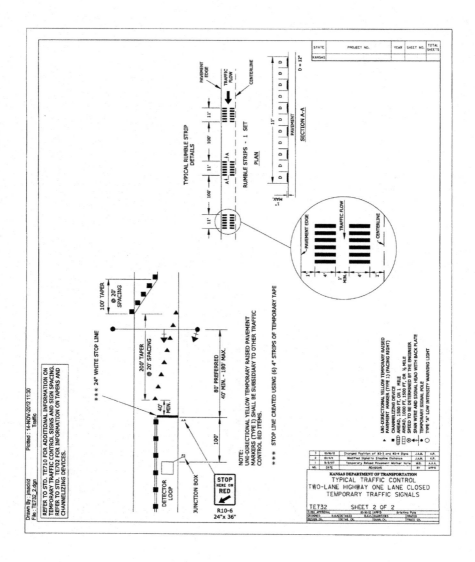

Work space = 500 ft., assuming a 300 ft. span bridge with 2–33 ft. approaches and staging areas either side.

The stop bar to stop bar maximum distance is: 180 ft. + 100 ft. + 165 ft. + 500 ft. + 100 ft. + 200 ft. + 180 ft. = 1425 ft.

Assume the highway speed limit is 65 mph and the work zone speed limit is 45 mph.

STEP 2

Check minimum time necessary for a semitrailer to accelerate from a stop and to clear the work zone.

From Figure 3.6, p. 65 of the 5th Edition of the *Institute of Transportation Engineers (ITE) Traffic Engineering Handbook*, the acceleration for a semitrailer to 45 mph is 1.1 ft./s².

For uniform acceleration from an initial velocity of zero (where l = length):

$$t = \frac{\sqrt{2al}}{a} = \frac{\sqrt{2(1.1 \text{ ft./s}^2)1425 \text{ ft.}}}{1.1 \text{ ft./s}^2} = 51 \text{ s.}$$

Assume a signal timing of 75 s, this would allow a truck to clear and allow eight other vehicles in a queue, assuming 2.5 s headways.

In conversation with Kristine Pyle, P.E. KDOT Work Zone Traffic Engineer, it was related that a signal timing of 60 s is usually used initially; but is adjusted by the field to meet local conditions.

STEP 3

Assuming that the signalized lane is being utilized under its capacity, assume a uniform delay and an effective green time of half the cycle.

$$d_c = \frac{1}{2}C\frac{\left(1-(g/C)\right)^2}{\left(1-(v/s)\right)} = \frac{75}{2}\frac{\left(1-(1/2)\right)^2}{\left(1-(393/1400)\right)} = 13.0 \text{ s}$$

The value 393 for v is the largest peak hourly volume for the signalized bridge projects in FY 2003. A value of 200 for v results in a uniform delay of 11.2 s.

Use 13 s as the uniform controller delay for the analysis.

STEP 4

Referring to the work zone layout in TE732, calculate the following travel times through the work zone for the four cases outlined. Assume the posted speed in the work zone is 45 mph and the highway speed is 65 mph.

(Note: 65 mph = 95.33 fps and 45 mph = 66.0 fps)

Case 1 – Passenger car travels through with no stop

1. Assume uniform deceleration from the initial warning sign to the speed sign, just past the first rumble strips:

Distance $= C + 2.5$ ft. $B = 750$ ft. $+ 2.5 \times 750$ ft. $= 2625$ ft.

$$\text{Time} = \frac{2l}{v + v_0} = \frac{2 \times 2265 \text{ ft.}}{95.3 \text{ fps} + 66 \text{ fps}} = 32.5 \text{ s.}$$

2. Travel at 45 mph from the speed sign to the far stop bar:

Distance $= A + B/2 + 1425$ ft. $= 750$ ft. $+ 750$ ft./2 $+ 1425$ ft.
$= 2550$ ft.

Time $= l/v = 2550$ ft./66 fps $= 38.6$ s.

3. Assume uniform acceleration from 45–65 mph:

$$\text{Distance} = \frac{v^2 - v_0^2}{2a} = \frac{(95.33 \text{ fps})^2 - (66 \text{ fps})^2}{2 \times 3.5 \text{ ft./s}^2} = 676.1 \text{ ft.}$$ where from Figure 3.7, p. 66, *ITE Traffic Engineering Handbook*, 5th Edition 3.5 ft./s^2 is the typical rate of acceleration for a passenger car to 45 mph.

$$\text{Time} = \frac{v - v_0}{a} = \frac{95.33 \text{ fps} - 66 \text{ fps}}{3.5 \text{ ft./s}^2} = 8.4 \text{ s.}$$

4. Calculate delay with respect to through travel at 65 mph:

Σ Distance $= 2625$ ft. $+ 2550$ ft. $+ 676.1$ ft. $= 5851.1$ ft.
Σ Time $= 32.5$ s $+ 38.6$ s $+ 8.4$ s $= 79.5$ s.

Time through at 65 mph $= 5851$ ft./95.33 fps $= 61.4$ s.
Delay $= 79.5$ s $- 62.4$ s $= 18.4$ s.

Case 2 – Semitrailer travels through with no stop

1. Assume uniform deceleration from the initial warning sign to the speed sign, just past the first rumble strips:

Distance $= C + 2.5 B = 750$ ft. $+ 2.5 \times 750$ ft. $= 2625$ ft.

$$\text{Time} = \frac{2l}{v + v_0} = \frac{2 \times 2265 \text{ ft.}}{95.3 \text{ fps} + 66 \text{ fps}} = 32.5 \text{ s.}$$

2. Travel at 45 mph from the speed sign to the far stop bar:

Distance $= A + B/2 + 1425$ ft. $= 750$ ft. $+ 750$ ft./2 $+ 1425$ ft.
$= 2550$ ft.

Time $= l/v = 2550$ ft./66 fps $= 38.6$ s.

3. Assume uniform acceleration from 45–65 mph:

$$\text{Distance} = \frac{v^2 - v_0^2}{2a} = \frac{\left(95.33\,\text{fps}\right)^2 - (66\,\text{fps})^2}{2 \times 1.1\,\text{ft./s}^2} = 2151.1\,\text{ft.}$$

$$\text{Time} = \frac{v - v_0}{a} = \frac{95.33\,\text{fps} - 66\,\text{fps}}{1.1\,\text{ft./s}^2} = 26.7\,\text{s}$$

4. Calculate delay with respect to through travel at 65 mph:

Σ distance $= 2625\,\text{ft.} + 2550\,\text{ft.} + 2151.1\,\text{ft.} = 7326.1\,\text{ft.}$
Σ time $= 32.5\,\text{s} + 38.6\,\text{s} + 26.7\,\text{s} = 97.8\,\text{s.}$

Time through at 65 mph $= 7326.1\,\text{ft.}/95.33\,\text{fps} = 76.8\,\text{s.}$
Delay $= 97.8\,\text{s} - 76.8\,\text{s} = 21.0\,\text{s.}$

Case 3 – Passenger car with stop

1. Assume uniform deceleration from the initial warning sign to the speed sign, just past the first rumble strips:

Distance $= C + 2.5\ B = 750\,\text{ft.} + 2.5 \times 750\,\text{ft.} = 2625\,\text{ft.}$

$$\text{Time} = \frac{2l}{v + v_0} = \frac{2 \times 2265\,\text{ft.}}{95.3\,\text{fps} + 66\,\text{fps}} = 32.5\,\text{s}$$

2. Assume uniform deceleration from the speed sign at 45 mph to the near stop bar:

Distance $= A + B/2 = 750\,\text{ft.} + 750\,\text{ft.}/2 = 1125\,\text{ft.}$

$$\text{Time} = \frac{2l}{v + v_0} = \frac{2 \times 1125\,\text{ft.}}{66\,\text{fps} + 0\,\text{fps}} = 34.1\,\text{s.}$$

3. Assume uniform acceleration from 0–45 mph:

$$\text{Distance} = \frac{v^2 - v_0^2}{2a} = \frac{\left(66\,\text{fps}\right)^2 - (0\,\text{fps})^2}{2 \times 3.5\,\text{ft./s}^2} = 622.3\,\text{ft.}$$

$$\text{Time} = \frac{v - v_0}{a} = \frac{66\,\text{fps} - 0\,\text{fps}}{3.5\,\text{ft./s}^2} = 18.9\,\text{s}$$

4. Travel at 45 mph to the far stop bar:

Distance $1425\,\text{ft.} - 622.3\,\text{ft.} = 1425\,\text{ft.} - 622.3\,\text{ft.} = 802.7\,\text{ft.}$
Time $= l/v = 802.7\,\text{ft.}/66\,\text{fps} = 12.2\,\text{s.}$

5. Assume uniform acceleration from 45–65 mph:

$$\text{Distance} = \frac{v^2 - v_0^2}{2a} = \frac{\left(95.33\,\text{fps}\right)^2 - (66\,\text{fps})^2}{2 \times 3.5\,\text{ft./s}^2} = 676.1\,\text{ft.}$$

$$\text{Time} = \frac{v - v_0}{a} = \frac{95.33\,\text{fps} - 66\,\text{fps}}{3.5 \times \text{ft./s}^2} = 18.9\,\text{s.}$$

6. Calculate delay with respect to through travel at 65 mph:

$\Sigma\text{Distance} = 2625\,\text{ft.} + 1125\,\text{ft.} + 622.3\,\text{ft.} + 802.7\,\text{ft.} + 676.1\,\text{ft.}$
$\qquad\qquad = 5851.1\,\text{ft.}$

$\Sigma\text{Time} = 32.5\,\text{s} + 34.1\,\text{s} + 18.9\,\text{s} + 12.2\,\text{s} + 8.4\,\text{s} = 106.1\,\text{s.}$

Time through at 65 mph = 5851 ft./95.33 fps = 61.4 s.
Delay = 106.1 s − 61.4 s = 44.7 s.

Case 4 – Semitrailer with stop

1. Assume uniform deceleration from the initial warning sign to the speed sign, just past the first rumble strips:

Distance = C + 2.5 B = 750 ft. + 2.5 × 750 ft. = 2625 ft.

$$\text{Time} = \frac{2l}{v + v_0} = \frac{2 \times 2265\,\text{ft.}}{95.3\,\text{fps} + 66\,\text{fps}} = 32.5\,\text{s.}$$

2. Assume uniform deceleration from the speed sign at 45 mph to the near stop bar:

Distance = A + B/2 = 750 ft. + 750 ft./2 = 1125 ft.

$$\text{Time} = \frac{2l}{v + v_0} = \frac{2 \times 1125\,\text{ft.}}{66\,\text{fps} + 0\,\text{fps}} = 34.1\,\text{s.}$$

3. Assume uniform acceleration from 0–45 mph:

$$\text{Distance} = \frac{v^2 - v_0^2}{2a} = \frac{\left(66\,\text{fps}\right)^2 - (0\,\text{fps})^2}{2 \times 1.1\,\text{ft./s}^2} = 1980\,\text{ft.}$$

$$\text{Time} = \frac{v - v_0}{a} = \frac{66\,\text{fps} - 0\,\text{fps}}{1.1\,\text{ft./s}^2} = 60.0\,\text{s.}$$

4. This is past the stop bar, so continue uniform acceleration from 45–65 mph:

$$\text{Distance} = \frac{v^2 - v_0^2}{2a} = \frac{(95.33\,\text{fps})^2 - (66\,\text{fps})^2}{2 \times 1.1\ \text{ft./s}^2} = 2151.1\,\text{ft.}$$

$$\text{Time} = \frac{v - v_0}{a} = \frac{95.33\,\text{fps} - 66\,\text{fps}}{1.1\ \text{ft./s}^2} = 26.7\,\text{s.}$$

5. Calculate delay with respect to through travel at 65 mph:

$\Sigma \text{Distance} = 2625\,\text{ft.} + 1125\,\text{ft.} + 1980\,\text{ft.} + 2151.1\,\text{ft.} = 7881.1\,\text{ft.}$
$\Sigma \text{Time} = 32.5\,\text{s} + 34.1\,\text{s} + 60.0\,\text{s} + 29.7\,\text{s} = 153.3\,\text{s.}$

Time through at 65 mph $= 5851$ ft./95.33 fps $= 82.7$ s.
Delay $= 153.3$ s $- 82.7$ s $= 70.6$ s.

STEP 5

Due to low v/c for the study bridges, ignore overflow and random delays. Assume half of arrivals arrive at green and half at red.

$$\text{Average delay} = \text{Delay}_{\text{controller}} + \text{Delay}_{\text{travel}}$$
$$= \text{Delay}_{\text{controller}} + \tfrac{1}{2}\,\text{Delay}_{\text{travel-no stop}} + \tfrac{1}{2}\,\text{Delay}_{\text{travel-stop}}$$

- For passenger cars: average delay $= 13.7$ s $+ 0.5 \times 17.1$ s $+ 0.5 \times 44.6$ s $= 44.6$ s.
- For semitrailers: average delay $= 13.7$ s $+ 0.5 \times 20.9$ s $+ 0.5 \times 70.6$ s $= 59.5$ s.

APPENDIX 2

Economic Impact Analysis of Deferring Maintenance on K-10 Bridges Near Desoto, Kansas

Federal Highway Administration (FHWA) defines economic impact analysis as "The study of the way in which the direct benefits and costs of a highway project (such as travel time saving) affect the local, regional, or national economy." It differs from benefit/cost analysis in that the subjects of benefit/cost analysis are the direct benefits and costs that a project has for highway owners and users; and concerning externalities, nonusers affected by the project. Economic impact analysis attempts to measure how these direct benefits and costs are converted into indirect effects in the economy.

Economic impact analysis for the Kansas Department of Transportation (KDOT) is typically performed using the web-based Transportation Economic Development Impact System (TREDIS). TREDIS consists of a series of modules maintained by the TREDIS Software Group, a division of the Economic Development Research Group, Inc. of Massachusetts. The system is used by approximately 20 states and provinces and assorted municipalities and federal agencies to conduct benefit/cost analysis and economic impact analysis for projects across multiple transportation modes.

The system can be used to assess user benefits based on transportation forecasting results and to calculate wider economic benefits based on the estimated impacts to jobs, income, gross regional product, and business output. What makes the system particularly attractive is the large pool of contemporary economic data that supporters use for their economic and transportation models.

To provide some measure of the impact of deferring maintenance on bridges specific to the scope of this report, a TREDIS analysis was conducted on a hypothetical scenario that assumes a pair of bridges on the Kansas State Highway System were allowed to deteriorate to a posted condition for a 2-year period before being repaired. The analysis was conducted by David Schwartz, P.E. and Andrew Jenkins, P.E. of the KDOT planning section.

The most minimal posting is one that restricts the bridge to only "legal" loads, those that do not require an Overweight permit issued by the Kansas Department of Revenue (KDOR). This would limit a non-Interstate

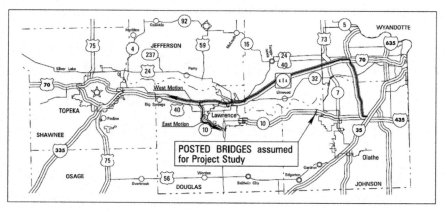

Figure A2.1 *Detour Overweight Truck Route for K-10 Posting.*

highway bridge to trucks with a gross weight less than 85,500 pounds. The annual Overweight permit issued by KDOR for $150 would allow a truck of up to 120,000 pounds to travel on any unrestricted bridges in the State of Kansas at will.

For this analysis, it was chosen to assume that at least one eastbound and one westbound bridge of the group of bridges repaired on K-10 near DeSoto, KS were posted "legal" for the years 2014 and 2015 (Figure A2.1).

It was assumed that trips with destinations west of Lawrence would detour on K-7 to I-70. For trips with destinations inside of Lawrence the detour was assumed to be K-7 to I-70 to K-10 around the west side of Lawrence. This was assumed since the two turnpike entrances to Lawrence have size restrictions that would prevent oversized trucks from using these routes. It was assumed for this study that only trips wanting to go to Lawrence and trips wanting to go west on I-70 via K-10 as a shorter route than K-7 would use K-10. Trips going south to Ottawa and beyond would use I-35 as a direct route and trips going north of Kansas City would take routes other than K-10 as they would be more direct, as well.

The volume of overweight trucks was determined from KDOT weight and motion studies averaged for all like roads to K-10 in the period 2003–2011. Superload (trucks over 150,000 pounds) permit data specifically for the K-10 bridges was averaged over years 2000–2006 and added in addition to the overweight vehicles.

Fifteen percent of overweight trucks and 5% of Superloads were assumed to have origins – destinations in Lawrence with the rest going west

Table A2.1 Travel impacts to oversized freight traffic.

Increases in	Truck freight
Gross vehicle trips	0
Gross VMT	535,297
Gross VHT	5,406
Gross buffer time (h)	135
Freight ton miles	37,474,318
Fatalities	0
Personal injuries	0.05
Property damage	0.89
Local portion of trip ends	7%
Total value of travel impacts	
Passenger cost – net total	0
Crew cost – net total	155,058
Freight cost – net total	456,844
Reliability cost – net total	8,427
Vehicle operation cost – net total	656,272
Toll cost – net total	318,952
Safety cost – net total	17,816
Environmental cost – net total	28,371
Induced benefit – total	0
Total value of travel impacts	1,641,740

Operation period: Start year, 2014; end year, 2015; analysis year, 2015; travel growth rate, 1.5%; constant dollar year, 2012.
© 2012 Economic Development Research Group, Inc.

on I-70. Tolls were taken into account to the Lecompton exit on I-70 for the detour routes.

Trips, VMT, and VHT were calculated and inputted into TREDIS along with other parameters. An analysis concerning overweight freight traffic only was conducted to determine the costs related to maintaining the detour for calendar years 2014 and 2015. The results are shown in Tables A2.1 and A2.2.

The manner in which the TREDIS analysis was conducted was to assume that the baseline condition for overweight freight traffic was the detour routes, then to determine the cost impacts from constructing the existing K-10 alignment as "new" construction. That should have provided a correct measure for adverse travel utilized for the user costs calculations shown in Table A2.1. The economic impact due to increased user costs on overweight freight traffic from posting K-10 near DeSoto was estimated to be $1.6 million in FY 2012 dollars.

Table A2.2 Economic impact related to oversized freight traffic

		TREDIS report: total economic impacts – by year			
Count	Year	Business output ($ million)	Value added ($ million)	Jobs	Wage income ($ million)
1	2014	−0.136	−0.063	0	−0.043
2	2015	−0.138	−0.064	0	−0.044
	Sum of impact for all years	−0.274	−0.127		−0.087

With regard to the total economic impact to the study area as reported in Table A2.2, it is not as clear that an economic benefit resulting from a transportation network improvement is the same value as the economic costs resulting from a degradation of the transportation network. However, the results are useful to compare the magnitude of total economic impact to those of user impacts. The loss of wages and output to the economy is nominally $488,000 for the 2-year period, compared to $1.6 million in user costs, for impacts on overweight freight movements.

SUBJECT INDEX

A

Abutments, 39, 72
  bearing seat on wall, 40
  continuity with superstructure, 40
  on deep foundation elements, 43
  integral with straight wings on girder
    bridge, 41
  K value, 248
  shelf-type, 41
  tilted finger, 39
  types, 42
  typical al plan of GRS, 43
Accuracy
  in load-rating analysis, 150
  for project maintenance, 281
ADT. *See* Average daily traffic (ADT)
Air–arc method, 196
Air peening, 196
A-Jacks™, 244
Alkali–silica reaction, 82
Aluminum, 34, 48
  rail, 49
American Association of State Highway
    and Transportation Officials
    (AASHTO), 295
  *AASHTO Guide for Commonly
    Recognized (CoRe) Structural
    Elements*, 117
  AASHTO LRFD Bridge Design
    Specifications, 3, 152, 196,
    202, 221
  AASHTO LRFD design truck, 6
  AASHTO *Manual for Bridge
    Evaluation*, 151
  AASHTO *Road and Bridge Maintenance
    Manual*, 155, 188
  AASHTOWare Bridge Management, 296
  condition states for reinforced concrete
    decks from, 120
  define a bridge management system, 110
  HL-93 design truck, 7
  recommend routine wading
    investigation, 103
  S – N curves used in specifications, 76

  spall development from cracking, 165
  states define legal trucks, 148
American Association of State Highway
    Officials (AASHO) code, 6,
    184, 237
Approaches, and roadways, 173
  approach settlement, 176–177
    concrete pavement expansion, 178
    options, to alleviate situation, 177
    pumping grout, into underlying
      void, 177
    settlement of approach slab, 176
  driver guidance, 173–176
  relief joints, 178
Arch, 62
  rib, 63
  shapes, 35
  steel cables for network arch, 88
  stone and masonry, 62
  use of reinforced concrete or steel, 62
Arkansas River, 19
Asset management, 296–298
  accountability and feedback, 297
  options/tradeoffs, analysis of, 297
  performance-based, 297
  policy driven, 297
  quality information, decisions based, 297
Automobiles, 9, 13, 289, 290
  traffic, 27, 31, 143
Average daily traffic (ADT), 111,
    256, 257
AWS D1.5 guidelines, 196
Axial compression, 69
Axial tension, 66

B

Backfall reconstruction
  as semi-integral abutment, 227
Bauschinger effect, 207
Beams, 42, 57, 71, 72, 74, 215, 241
  bridge, superstructure components
    on, 48
  fabricated from, 58
  stresses, effect of tensioning, 186

Bearings, 53–56, 226
connection, 206
elastomeric, 55
reinforced, 56
pot, 55, 56
steel type, 54
Bedrock material, 35
Bending, 70
moment in a beam, 71
stress, 71
Bid-price index (BPI), 261, 262
Bituminous overlay, 217
BLCCA. *See* Bridge life cycle cost analysis
(BLCCA)
BMS. *See* Bridge management system
(BMS)
Box structures, 37
BPI. *See* Bid-price index (BPI)
Bridge collapse
I-35W bridge, 25, 28
in the United States, 212, 244
Bridge components, 259
Bridge deck, 5
bituminous overlay, 218
Bridge deck repair work, undercapacity
flow calculation
average uniform delay from signal
controller, 311, 315
2010 Highway Capacity Manual, 311
highway speed limit, 314
KDOT. *See* Kansas Department of
Transportation (KDOT)
signalized work zones review for bridge
deck repair, 311
total average delay for passenger car/
semitrailer, 312, 319
traveling delay calculation through work
zone, cases, 311, 315–319
passenger car travels with no stop,
315–316
passenger car with stop, 317–318
semitrailer travels with no stop,
316–317
semitrailer with stop, 318–319
typical layout of work zone
determination, 311, 312
typical signal timing determination,
311, 315

Bridge decks, and expansion joints, 160
deck and crack sealing, 165–167
deck sealing operation, 167
properties of concrete sealers, 166
spall development from cracking, 165
deck and expansion joint washing,
167–168
deck drainage, 160
drain extension retrofit, 162
flush truck in operation, 161
plugged deck drain, 160
rock flume-off of bridge
approach, 161
deck patching, 162–164
area prepared for partial depth
patching, 164
deck deterioration at expansion
joint, 163
drain tube through timber planks, 164
timber deck preservation, 168
Bridge Deck Service Life Prediction and Cost,
report, 279
Bridge Element Deterioration Rates for
NYSDOT, report, 276
Bridge engineering community, 19
Bridge evaluation step, 5
Bridge failure, 11
documentation of, 10
partial list, report of KHC, 11
the United States, 2, 103
Bridge health index (BHI), 303, 304
screening tool and performance
measure, 305
State Highway System
priority ranking procedure, 305
TEV and CEV, 305
Bridge inspections, 14
Canada
Alberta *Bridge Inspection and
Maintenance Manual,* rating, 122
British Columbia, bridge, 122
Federal government, role in, 121
manuals to govern inspections
on, 121
Ontario Structure Inspection Manual,
122, 123
Quebec implemented, management
system, 123

Saskatchewan, major bridges inspected by, 122
Finland, 127
　defect ratings in, 128
　Finnish Road Administration, role for, 127
　inspector certification, 128
　repair index formula for bridges in, 128
France, 124
　cursory inspections, conducted annually to, 126
　French National Road Directorate, 124
　Inspection procedures, codified in, 126
　levels of certification for, 126
Germany, 126
　BMVBW, administered by, 126
　damage assessments for structural elements, 127
　levels of routine inspection, by bridge inspectors, 126
and maintenance, 14
under the NBIS, 27
observations, 129
　approaches to inspector requirements and certification, aspects, 129
　United States *vs.* other countries, in managing inspection aspects, 110–112
programs, 18
provide data for, 99
South Africa, 124
　DER ratings for inspections in, 125
United Kingdom, 123
　FHWA, scanning tour of bridge inspection, 123
　NCHRP report on, 123
　UK Bridges Board Group, tasked with developing standards and, 124
　United Kingdom Highways Agency, role on, 123
United States, 100
　appraisal ratings, 110–112
　component condition ratings, 105–109
　critical findings, 113–114
　damage inspections, 103
　deficiency and sufficiency, 112–113
　element-level inspection, 114–121
　FHWA component condition-rating guidelines, 106
　requirements to be a team leader for, 101
　structure inventory and appraisal sheet, 102
　techniques, 103
　types of inspection, 100, 104
　underwater inspections, 103
Bridge inspectors, 5, 204
Bridge life cycle cost analysis (BLCCA), 259, 301
and activity diagram, 265
annual average consumer price index, 260–262
bridge elements of NY bridges, 278
cash flow diagrams, 263–265
deck patching and polymer overlay, 268
deteriorated reinforcing steel in parapet, 283
deterioration rates, 275–281
　Analysis of Life Cycle Maintenance Strategies for Concrete Bridge Decks, 280
　Bridge Deck Service Life Prediction and Cost, 279
　Bridge Preservation Guide, 280
　Developing Deterioration Models for Nebraska Bridges, 280
　Development of Agency Maintenance, Repair & Rehabilitation (MR&R) Cost Data for Florida's Bridge Management System, 279
　Guidelines for Selection of Bridge Deck Overlays, Sealers and Treatment, 280
　plots for structural decks owned by NYSDOT, 278
determining costs, 266
　agency costs, 266–270
　economic costs, 275
　user costs, 270–274
　vulnerability costs, 275
discount rate, 262–263
element service lives, 281
FHWA average annual composite bid-price index, 261

Bridge life cycle cost analysis (BLCCA)
(*cont.*)
FHWA Life-Cycle Cost Analysis Primer
lists, 259
FHWA national highway construction
costs index, 262
inflation, 259
KDOT agency cost, 269
to bituminous patch, 269
inflation rates, 263
to seal, 268
maintenance effort
costs and condition, 258
for Munker's Creek Bridge, 285
NY State DOT condition
ratings, 276
overview, 255
programs, 282, 282c
project scoping/selection, 255
adjusting scope, 257–258
initial selection, 255–257
residual value, 265
sample cash flow diagram, 264
sample delay calculation, at work
zone, 273
simple condition diagram, 264
types of costs, 266
typical applications of, 283–286
Bridge maintenance
activities, 270
lowest agency cost, 270
system-wide review of, 258
Bridge Management Engineer for
KDOT, 284
Bridge management system (BMS)
asset management, 296–298
accountability and feedback, 297
options/tradeoffs, analysis of, 297
performance-based, 297
policy driven, 297
quality information, decisions
based, 297
basic modules, 294
bridge inspection report, 300
candidates for maintenance work, 301
case study of, 297
contemporary history of, 289
definition of, 291–293

FHWA Bridge Management
Systems, 298
flowchart, 291, 292
functionally obsolete (FO), 300
Kansas department of transportation, 308
commitment to bridge
preservation, 308
inspection/design, coordination, 309
level of service, 293
maintenance recommendation, 300
measures for bridge improvement
needs, 293
NBI rating, 300
before the NBIS, 289–291
NCHRP study, 296
NHS bridges in good/poor
condition, 298
optimization routines of, 298
overview, 289
preservation cycle of inspection, 293
priority index (PI), 298
project/program selection, 298
maintenance work, 299–302
priority indexing, 302–308
priority ranking, 293, 298–299
setting program priorities, prototypical
processes for, 299
software, 294–296
implementations, 295
structurally deficient (SD), 300
top-down approach, 296
truss bridge testing, moving loads, 290
Bridge Management Systems Software, 294
Bridge mechanics, 65
axial forces–compression, 68–70
axial forces–tension, 66–68
Bridge opening, 2
cross-section, 249
Bridge preservation, 3, 5–8, 27, 99
activitie involving, 5, 29
BMS systems, 27
development of BMS software, 28
funding scenarios on health of
network, 28
management, 27–28
PONTIS led to changes in, 28
practices, before 1970, 8–12
standardized NBIS format, 27

Bridge projects, 274
Bridges, maintenance activities, 290
Bridge steels, 210
Bridge structures, 32
    classification of, 31
Bridge substantial maintenance, 308
    funding, 308
    program manager for, 308
Bridge substructure, and waterway, 172
    drift accumulation, 172
    erosion under abutment, 175
    maintenance forces placing
        "A-Jacks", 174
    scour at pier wall, 173
    use of stone as revetment at berms and
        scour protection at, 175
Bridge superstructure, and substructure, 168
    bearing device maintenance, 170
    painting, 170–171
    sealing bearing seats, 169
    washing superstructures, 168
Buckling, 23, 24, 69, 70, 74, 92
    capacity, 209
Buried structures, 32–37
    foundations of, 35

C

Cable-supported bridges, 64
Carbon content, of steel, 89
Cash flow diagram, 263, 264
Cathodic protection methods, 191
Cementitious materials, 190
Center span, of I-35W before collapse, 20
Charpy impact test, 90
Chloride-contaminated concrete, 190
Chloride exposure, 82
Chloride-free concrete, 191
Christopher bond bridge–cable stayed, 65
Code of Federal Regulations, 291
Cold working steel, 207
Collapse in Minnesota, 19
Collapse of the I-35W Bridge, 19–22
Complex bridges, 64
Compression, 21, 68, 70, 72, 84, 86
    joint seals, 53
    member, 138, 209, 234
Compressive stress, 71, 72, 92, 196,
        197, 238

Concrete, 77
    bridges, prestressed, 240
    deterioration, 83
    epoxy material, selection, 187
    hydroxide ions, 190
    low-tensile strength of, 186
    matrix, 81
    patching, 188, 189
    in piers, 243
    piles, 46
    prebagged, 190
    prior to patching, 191
    repair methods. *See* Concrete, repair
        methods
    replacing deteriorated, 215
    spall, 83
    wingwall, 37
Concrete, repair methods, 185
    corrosion resistance, 190
    crack repair, 186
    drilled/grouted reinforcing, 192, 193
    material selection, 189
    patching/surface repair, 188, 189
    stresses, phasing/distribution of, 193
Connecticut Department of
        Transportation, 16
Construction costs, 267
Construction of prestressed concrete
        inverted-T beam bridge, 85
Contemporary structural steels, 89
Corral rail, 49. *See also* Railing
Corrosion, 25, 194. *See also* Rust
    post-tension strands, 242
    reinforced concrete, 191
    reinforcing steel, 190
    resistance, 89, 98, 190
    steel, 202
        reinforcing, 82
    through web of steel beam, 92
Corrugated steel, 34, 251
Cost, 1, 90, 225, 236, 258. *See also*
        Bridge life cycle cost analysis
        (BLCCA)
    determining, 266–275
Cost effectiveness, 156–157
Cost of failure, 26
    analysis by MnDOT, 27
    I-35W Bridge, 26

Cracks, 74, 79, 195
  geometries, 75
  growth, 75
    per cycle as a function of
      stress-intensity factor, 76
  repair, 186
  sealing, 240
  size, 74
  tip hole, location of, 197
Critical assets, 1
Crushing, 74
Culvert
  structures, 2, 33
Culverts, 32
  structures, 33
Cutting stiffener, to soften web
      gap, 201
Cyclic loading, 194
Cypress Street Viaduct, 19

D

Decks, 49
  deterioration in material, 50
  drains, 16, 52
    maintenance, 52
  expansion joints, 221
  filled steel grate, 51
  forming, for redeck project, 252
  joints, 53
  patching, 269
    project, 214
  reinforced concrete, 51
  timber, 51
  trusses, 24
Deformation, plastic, 207
  treatment of, 204
Delphi process, 305
Department of Transport (DOT), 257
Design codes, 184
  inherent in, 184
Design error, 23
Design specifications, 6
DeSoto Bridge, 24
Deteriorated steel rocker bearing, 229
Deterioration, 28, 91, 115, 155, 186, 219
  from an inherent deficiency, 84
  gusset plate on Truss Bridge, 25
  rates, and relevant studies, 275–281

Analysis of Life Cycle Maintenance
    Strategies for Concrete Bridge
    Decks, 280
Bridge Deck Service Life Prediction
    and Cost, 279
Bridge Preservation Guide, 280
Developing Deterioration Models for
    Nebraska Bridges, 280
Development of Agency Maintenance,
    Repair & Rehabilitation
    (MR&R) Cost Data for Florida's
    Bridge Management
    System, 279
Guidelines for Selection of Bridge
    Deck Overlays, Sealers and
    Treatment, 280
  plots for structural decks owned by
    NYSDOT, 278
Detouring traffic, 271
Development of Agency Maintenance,
    Repair & Rehabilitation
    (MR&R) Cost Data for Florida's
    Bridge Management
    System, 279
Disc bearings, 56
Distortion-induced damage, 152
  due to secondary stresses, 151
Distortion-induced fatigue, 198
  damage, 198
Drainage
  abutment system, 247
  deck, 52, 160
  uncontrolled, 247
Drill, 193, 202
  holes, 199
  shafts, 47

E

Earthquake, 19
Edge crack, in plate, 75
Elastic strain, 67
Elastomer, 54, 55, 77, 226
Elastomeric bearings, 227
Engineering analysis, 5
Epoxy injection, 187
Epoxy material, selection, 187
Erie Canal, 8
Euler buckling formula, 69

Evaluation, 5
    component-level, 109
    of existing bridge details for, 152
    fatigue, of steel bridges, 151–152
    structural, 110, 111, 113
Expansion joints, 53
    categories, 53

F

Fatigue, 32, 74
Fatigue cracking, 25, 185, 194
    bolted splice, 198
    in steel, 93
FDOT Contract BB-879, 279
the Federal Aid Highway Act of 1916, 14
Federal Emergency Management
        Agency, 212
the Federal Highway Act 1968, 13
Federal Highway Administration (FHWA)
    bridge, defined as, 1
    bridge opening, 2
    for bridges, actions being mandated
        by, 23
    for bridge work, 291
    guide *Heat Straightening Repairs of
        Damaged Steel Bridges*, 207,
        210, 271
    Highway Bridge programs, 303
    Life-Cycle Cost Analysis Primer, 259
    *Manual of Repair and Retrofit of Fatigue
        Cracks in Steel Bridges*, 195, 196,
        199, 200
    *Work Zone Road Users Costs-Concepts and
        Applications* report, 271
FHWA. *See* Federal Highway
        Administration (FHWA)
Fiberglass, 34
Fiber-reinforced polymer (FRP), 97
    concrete column repair, typical, 244
    reinforcing strips, 239
    systems, 240
        prestressed beam repair, 242
Fiber resin polymers, 52
    performance, 52
Fillet on reinforced concrete box
        culvert, 34
Finger plate expansion joint, 221
Fire damaged steel girders, 213

Forces from transverse loading, 70
Foundations, 45–47
    deep, 46
Fracture, 74
    critical members, 16
    critical structure, 21
    mechanics, 74
    toughness, 90
        of steel is affected by, 90
Freeze–thaw cycle, 83
Froude number, 248
FRP. *See* Fiber-reinforced polymer
        (FRP)
Functionally obsolete bridge, 183

G

Galvanic anode, 191, 192
Girder, 57
    fabricated from, 58
Gravity poured sealers, 186
Grinding, 195
    small nicks and gouges, 208
Gross vehicular weight (GVW), 6
Grout, 193
Gusset plates, 24, 25, 232
    deteriorated gusset plate, repair
        for, 233
    fatigue issues, 25
    repair, 26

H

Heat straightening, 210
    restraint, 211
Highway bridges, 1, 265
    challenge of structures, 181
    construction
        high/low-tensile capacity
            material, 185
    cost-effective approach, 181
    deck construction, 218
    deteriorate, 181
    substantial maintenance actions, 182
Highway capacity manual (HCM)
    Capacity Reductions due to
            Construction and Major
            Maintenance Operations of the
            2010, 272
HL-93 truck configuration, 6, 7

Horseshoe crack, at web gap, 200. *See also* Cracks

H15-S12 design truck, 7

HS20-44 truck configuration, 6

Hybrid web gap stiffening repair, 203

Hydraulic features, 32

Hydraulic roughness, 250

Hydrostatic pressure, 247

I

I-95 Bridge, 233

I-70 bridges, 322

Inlet end of reinforced concrete box, with soil saver, 36

Inspections, 5, 24

  by the KHC Bridge Maintenance Engineer, 12

Inspection techniques/technologies, 134

  inspecting fiber-reinforced polymer, 138–139

  nondestructive testing – concrete, 135–136

  nondestructive testing – steel, 136–137

  posttensioning ducts, 139–140

  sampling, 137–138

  sonar and underwater inspection, 142–143

  structural health monitoring (SHM), 140–142

  visual inspections and sounding, 134

Inspectors, bridge, 204

Integral abutments, 41

Intermodal Surface Transportation Efficiency Act (ISTEA), 296

I-35W Bridge, 19

  collapsed, 20

    causes of, 21

  connections at Node U10 West, 22

  East elevation of, 21

  failure of one of components of a deck truss, 21

  FHWA for bridges, actions being mandated by, 23

  locationation of Node U10, 23

  location of aggregate and equipment on, 22

  NTSB investigation, 21

    final report, 22

I-35W Bridge, 232

J

Jacking force limit, 211

Joints

  deck expansion, 233

  expansion joint characteristics, 225

  glands, 222

  glands, 222

  pin/hanger joint, replacement of, 235

K

Kansas Department of Revenue (KDOR), 321

Kansas Department of Transportation (KDOT), 262

  agency's performance in bridge management and preservation, 308

  bridge priority formula, 307

  economic impact analysis, 321

  inspecting bridges, 24

  performance goal, 305

  TE710 for Rural Highways, 312

  Traffic Engineering Standard TE700, 312

  Traffic Engineering Standard TE732, 311–314

  weight and motion studies, 322

Kansas road system, 9

Kansas State Highway Commission (KHC), 10

K-7 bridges, 322

K-10 bridges

  detour overweight truck route, 322

  economic impact to oversized freight traffic, 324

  Federal Highway Administration (FHWA), 321

  Kansas Department of Transportation (KDOT)

    economic impact analysis, 321

  near DeSoto, 323

  travel impacts to oversized freight traffic, 323

KDOR. *See* Kansas Department of Revenue (KDOR)

KDOT. *See* Kansas Department of Transportation (KDOT)

KHC. *See* Kansas State Highway Commission (KHC)

L

Load-capacity requirements, 9
Load-carrying capacity, 6, 46, 204
Load-carrying element, 214
Load-induced deterioration, 83
Load-induced fatigue cracks, 198
Load ratings, 6, 100, 143
    analysis, 25
    analysis methodologies, 144–147
        allowable stress equation, 145
        FHWA started with inception of
            NBIS, 146
        form of the load factor equation, 145
        LFD specifications, 146
        load and resistance factor design
            (LRFD), 146
        load rating analysis, review for design
            truck loading at, 146–147
        HL-93 Design Load, 147
    general approach, 144
        calculating RF for each load
            effect, 144
        general form of equation, 144
    mannual, guiding specification for, 143
    procedure, 26
    by testing, 150–151
    truck loadings, 148–150
        AASHTO Type 3 Legal Load
            Trucks, 148
        maximum gross vehicular weight
            (GVW), 148
        notional rating load, 150
        single unit SHV loadings, 149
        specialized hauling vehicles
            (SHV), 149
        states define legal trucks, 148
Long-range transportation plan
    (LRTP), 257
LRFD fatigue load, 200

M

Maintenance, 5, 7
    activities, costs for, 268
    goal of, 8
    inspections, 158–159
Manganese, 89
Map cracking, 79
Markov chains approach, 277

Masonry plate, 54
Membrane sealant, replace strip seal
    gland, 224
Metropolitan Planning Organizations
    (MPO), 257
the Mianus River Bridge, 15
    collapsed section of, 16
Micropile, installation of, 48
the Minnesota Department of
    Transportation (MnDOT), 24
    inspecting structures with deck
        trusses, 24
Mississippi River, 19
Modern highways, 31
Modular joint expansion joint, 223
Modulus of elasticity, 68
Moment of inertia, 70
the Moving Ahead for Progress in the 21st
    Century (MAP-21), 298
Multiple cell at-grade box bridge, 34
Multiple corrugate metal pipes, 35

N

National bridge inspection standards
    (NBIS), 13
    changes to, 18
    development of, 13–18
    sequence of failure in pin and hangar
        assembly per, 17
    update to, 18
National Bridge Inventory (NBI), 3, 240,
    256, 291
    condition ratings, 306
    inspection data, 275
National Highway Construction Cost
    Index (NHCCI), 262
National Highway System Designation
    Act, 296
National Transportation Safety Board
    (NTSB), 16
NCDOT bridge
    clear deck width goals for two lane
        routes, 294
    load capacity, 293
    vertical clearance goals, 294
NCHRP Project 12-28, 294
NCHRP Project No. 12-85, 212
NCHRP Project 20-07 Task 307, 242

NCHRP Report 271, 208, 209, 210
NCHRP Report 300, 294
NCHRP Report 483, 274, 275
NCHRP Report 551, 296
NCHRP Report 604, 212
NCHRP Report 655, 240
NCHRP Report 483 *Bridge Life-Cycle Cost Analysis*, 282
NDT methods, 243
Nebraska Department of Roads (NDOR), 301
Neoprene gland expansion joint, 224
Network tied arch bridge, 64
New York State Department of Transportation, inspecting bridge structure, 24
Nonstructural cracking, 79
NYSDOT data, 277, 278
NYSDOT elements, 276
NY State Thruway Over Schoharie, 18
    collapse of, 18

O

Oakland Bay Bridge, 19
Office of Management and Budget guidelines, 263
Offset measurements, calculate radius of curvature, 209
Ohio River, 13
Ongoing changes in practice, 13, 19
Overcoating systems, 236

P

Painting, 204
Paint removal, on steel girder bridge, 237
Paris' law, 75
Park Service, 14
Patching concrete
    primary mitigation for corrosion, 191
Pier bents, configurations, 45
Pier cap, posttension retrofit of, 239
Piers, 44
    and bent types, 44
    concrete, 243
    placing rock, 246
    rock riprap placement, 246
Pile groups
    stiffness of, 226

Pipe structures, 37
Plastic, 34
Plastic deformations, 207
    repair of, 208
    treatment of, 204
Plastic shrinkage cracking, 81
Plastic strain, 67
Plate repair, angle and backing, 204
Polymer membrane, 8
Portland Cement Concrete, 189
Posttension ducts in Haunched slab, 85
Post-tensioning
    retrofit for, 238
posttensioning anchorage, 243
Pre-NBIS Kansas bridge inspection form, 12
Pressure relief joints, 179
Prestress concrete beam
    FRP repair installation on, 239
Prestressed concrete, 45, 84
    beam bridge, 58
    beam with FRP
        repair of, 98
    bridges, 240
    deterioration, 86
    general usage, 84–85
    properties, 86
Prestressing, effect of, 84
Prestressing strand, mechanical splice for, 241
Preventative maintenance, 8, 155
    AASHTO Subcommittee on Maintenance define as, 155
    categories, as per AASHTO Maintenance Manual for, 155
    goals for, 156
    mitigating threats, 156
    recommendations, 178
Price, 1
Private toll facility, 13
Programming maintenance actions, 152
PTFE sheet, 227
Public road facilities, 31
    categories, 31
    geometry of facility, 31

R

Railing, 47
    damaged type, 50

RAND Corporation, 305
RealCost, 282
Rehabilitation, 8, 182, 251
  bridge widening, 253
  deck replacement, 251–252
  funds, 2
  superstructure replacement, 253
Reinforced concrete, 45, 48, 77, 82
  corrosion, 191
  deterioration, 78
  general usage, 77
  properties, 77–78
  smaller diameter, shaft, 47
Reinforced concrete arch, 35
  bridge, 63
  culvert line, 251
Reinforced concrete box
  with concrete apron, 36
  culvert, 33
Reinforced concrete girder, 238
Reinforced elastomeric bearing, splitting
    of, 229
Reinforcing steel, 47, 191
Reliability-based bridge inspection, 129
  implementation in Indiana, 131–133
  risk-based assessment, 129–131
Restore capacity, 187
Rigid frames, 60
Riprap, at abutment, 250
Roadway width, 31
Rock armoring, 244
Routing, crack treated, 187
Rust, 24, 91, 120, 191, 202

S

Safe-load-carrying capacity, 15
Scaling, 82
Scour, 248
Sealed joints, 222
Sealing
  cracks, 186
    treated, 187
Section loss
  inside of truss chord, 205
Shear, 72
Shear crack, in concrete girder, 80
Shear force, 73
Shear near bearing, 73

Shear strengthening
  rebar insertion, 238
Shear stresses, 73–74
Short span timber bridge, 94
Shrinkage cracking, 81
Shrinkage cracks, 81
Silicon, 89
Silver Bridge, 13
  wreckage on the Ohio Side, 14
Silver Point Bridge, 13
Slab/overhang repair, 217
Slab/rail replacement, edge of, 220
Sliding plate expansion joint, 221
Slip critical connection, 206
S–N curves, used in AASHTO
    specifications, 76
Soften web gap
  cutting stiffener, 201
  large hole retrofit, 201
Soil saver, 36
Spalling, 83
Span bridge structures, elements of, 38
Span highway bridge, typical parts of, 39
Span structures, 56
Steel, 48
  cold working, 207
  corrosion, 202
  deterioration, 91–92
  fabrication shops, 207
  general usage, 86–87
  properties, 87–90
  reinforcing, 191
  repair methods, 194
    corrosion repairs, 202–206
    fatigue repairs, 194–202
    fire damage, 212–213
    impact repairs, 207–212
Steel beam bridge, 220
Steel beam, posttensioning of, 234
Steel bearings, 54, 226
  configurations, 54
  removal and reset, 231
Steel bolster
  parts of, 228
Steel cables for network arch, 88
Steel corrosion, 24. *See also* Corrosion
  protective coatings, 236
Steel erection, 233

Steel girder bridge, 58
  typical framing components for, 59
Steel girders, fire damaged, 213
Steel girder spans, 59
Steel girder superstructure, 253
Steel grid deck, repair of, 219
Steel H-piles, 46
Steel pier beam on straddle bent, 45
Steel pile bent, 87
Steel rocker, parts of, 228
Steel superstructures
  retrofit for, 238
Steel truss bridge, 32
  adding section to built-up member, 234
  deteriorated gusset plate on, 233
Stone arch bridge, 63
Stone masonry, 96
  substructure, 97
Straight and flared RCB wings, 38
Strain, 66
Stress corrosion cracking, 13
Stresses for timber construction,
    comparative allowable, 95
Stress-intensity factor, 74, 75, 194
Stress–strain diagram, 67
Stringer and floorbeam system, 59
Strip seal expansion joint, 222
Structural analysis, 6
Structural sab, 57
Substantial maintenance, 8, 182, 213
  abutments, 247–249
  bearing devices, 226–231
  bridges. See Substantial maintenance of
      bridges
  culverts, 249–251
  decks/railing, 214–219
  expansion joints, 221–226
  painting steel structures, 236–237
  piers/pier bents, 243–247
  posttensioned concrete superstructure,
      242–243
  prestressed concrete superstructure,
      240–242
  reinforced concrete superstructure,
      237–240
  steel superstructure, 231–235
Substantial maintenance of bridges
  assessment/scoping, 182–185

  closing/removing bridges, 182–183
  design codes and specifications,
      184–185
  level of repair, 183–184
  defined, 181
  repair methods, 185
    concrete, 185
      corrosion resistance, 190
      crack repair, 186
      drilled/grouted reinforcing,
          192, 193
      material selection, 189
      patching/surface repair, 188, 189
      stresses, phasing/distribution of, 193
    steel, 194
      corrosion repairs, 202–206
      fatigue repairs, 194–202
      fire damage, 212–213
      impact repairs, 207–212
Superstructures, 47
Surface Transportation and Uniform
      Relocation Assistance Act of
      1987, 15
Surface Transportation Assistance Act
      of 1978, 15
Surface Transportation Assistance Act of
      1982, 252

T
Teflon sliding
  From PTFE Bearing, 230
Temperature cracks, 80
Temporary wingwall, 37
Tensile force, 66
Tensile stress, 72, 74
Thin-bonded overlay, 217
Timber, 92
  deterioration, 95–96
  general usage, 92–94
  properties, 95
Timber abutment bent, 94
Timber deck, 51
Timber, defects in, 95
Timber headwall on corrugated metal pipe,
      37
Torque, 70
Torsion, on beam, 71
Track deterioration, 5

Traffic, 1
   control measures, 182, 269
Traffic loading
   and reactions on span bridge, 65
Transportation Economic Development
         Impact System (TREDIS), 321
   hypothetical scenario, 321
   Software Group, 321
   trips, VMT, and VHT, 323
Transportation improvement plan
         (TIP), 257
Transverse loadings, 72
Traveling delay calculation through work
         zone, cases, 311, 315–319
   passenger car travels with no stop,
         315–316
   passenger car with stop, 317–318
   semitrailer travels with no stop, 316–317
   semitrailer with stop, 318–319
Triaxial constraint condition on Hoan
         bridge, 91
Truss chord
   repair splice, 206
   section loss, 205
Trusses, 61, 232
   bridge configurations, 61
   components, 62
Turner–Fairbank Research Center, 196

U

Ultrasonic impact treatment (UIT), 196
Ultrasonic thickness gages, 24
Underwater components, of bridges, 18
United States
   amount expended by, 2
   bridge deterioration, 3, 4
   bridge failures, 3
   bridge inspection in, 100
      appraisal ratings, 110–112
      component condition ratings, 105–109
      critical findings, 113–114
      damage inspections, 103
      deficiency and sufficiency, 112–113
      element-level inspection, 114–121
      requirements to be a team leader
            for, 101
      structure inventory and appraisal
            sheet, 102

techniques, 103
      types of inspection, 100, 104
      underwater inspections, 103
   bridge on the NBI, 4
      deficient bridges, 4
      functionally obsolete, 4
   bridge opening, 2
   bridge preservation, 3
   bridges, 1
   bridges by decade of construction, 4
   bridges constructed, post-World War II
         period, 3
   collapse of a bridge, 2–3
   cost of bridges, 2
   FHWA statistics for road and
         bridges, 1
   length of bridges, 2
   load-carrying capacity, 4
   maintenance issues, 2
   rehabilitation funds, 2
   wearing surface, 3
United States Bureau of Labor
         Statistics, 259
User-cost calculations, 274

V

Vee heat pattern, 210
Vehicle operation costs (VOC)
   calculations for, 271
   for Kansas, 274

W

Waterproofing cracks, 186
Wearing surface of deck
   with polymer overlay, 50
Wearing surface of extruded FRP deck
         panels, 52
Weathering steel, 90
Web gap distress, 199
Web gap stiffening repair, 203
Weibull distribution approach, 276, 277
Welded steel rigid frame bridge, 60
Welding, 206
West Virginia Bridge Maintenance
         Engineer, 14

Z

Zinc-rich system, 236